Collins

Exploring Science | Grade 9

T0173425

Derek McMonagle

Reviewers: Marlene Grey-Tomlinson, Bernadette Ranglin, Maxine McFarlane & Monacia Williams

Collins

William Collins' dream of knowledge for all began with the publication of his first book in 1819. A self-educated mill worker, he not only enriched millions of lives, but also founded a flourishing publishing house. Today, staying true to this spirit, Collins books are packed with inspiration, innovation and practical expertise. They place you at the centre of a world of possibility and give you exactly what you need to explore it.

Collins. Freedom to teach.

Published by Collins
An imprint of HarperCollins*Publishers*
The News Building
1 London Bridge Street
London
SE1 9GF

Browse the complete Collins Caribbean catalogue at
www.collins.co.uk/caribbeanschools

10 9 8 7 6 5 4

ISBN 978-0-00-826329-4

British Library Cataloguing in Publication Data
A catalogue record for this publication is available from the British Library.

Publisher: Elaine Higgleton
Commissioning editor: Tom Hardy
In-house senior editor: Julianna Dunn
Author: Derek McMonagle
Reviewers: Marlene Grey-Tomlinson, Bernadette Ranglin, Maxine McFarlane & Monacia Williams
Project manager: Alissa McWhinnie, QBS Learning
Copyeditors: David Hemsley & Mitch Fitton
Proofreaders: David Hemsley & Helen Bleck
Photo researcher, illustrator & typesetter: QBS Learning
Cover designer: Gordon MacGilp
Series designer: Kevin Robbins
Cover photo: blew_s/Shutterstock
Production controller: Tina Paul
Printed and bound by: Grafica Veneta SpA in Italy

MIX
Paper from
responsible sources
FSC™ C007454

This book is produced from independently certified FSC™ paper to ensure responsible forest management.

For more information visit: www.harpercollins.co.uk/green

Contents

Introduction – How to use this book   4

Unit 1:   Working like a scientist   6

Unit 2:   Transport across cells   26

Unit 3:   Transport in humans   38

Unit 4:   Transport in plants   54

Unit 5:   Static electricity   68

Unit 6:   Current electricity   80

Unit 7:   Magnetism   108

Unit 8:   Chemical bonding, formulae and equations   130

Unit 9:   Sensitivity and coordination   154

Unit 10:   Acids and alkalis   174

Unit 11:   Embryo development and birth control   202

Index   222

Acknowledgements   229

Introduction – How to use this book

Unit 2: Transport across cells

We are learning how to:
- explain the movement of particles during diffusion
- relate diffusion to concentration gradient.

This tells you what you will be learning about in this lesson.

Diffusion

You already know something about the process of **diffusion**, although you may never have heard it called by this name. Living things need to move substances around. One way in which this is done is by diffusion. Sometimes when you step into your home, you know that someone has been baking or cooking.

When someone is baking or cooking, tiny particles of food move through the air. The particles travel in the air from the kitchen to all the other rooms in your home. We call this spreading out of particles 'diffusion'.

FIG 2.1.1 Jerk chicken

This introduces the topic.

The book has plenty of good illustrations to put the science into context.

Activity 2.1.1

Investigating diffusion

When ammonia gas and hydrogen chloride gas react they form a fine white 'smoke' of ammonium chloride particles. Your teacher will show you.

Here is what you need:
- Long glass tube with bungs at either end
- Cotton wool
- Concentrated aqueous ammonia
- Concentrated hydrochloric acid
- Stand and clamp.

Here is what you should do:
1. Support the tube horizontally using a stand and clamp.
2. To produce hydrogen chloride gas, wet a plug of cotton wool with hydrochloric acid, place it at one end of the tube and seal it with a bung.
3. To produce ammonia gas, wet a plug of cotton wool with aqueous ammonia, place it at the opposite end of the tube and seal it with a bung.

Each spread has activities to help you to investigate the topic.

There are often some fascinating fun facts.

4. What do you observe inside the tube after a short time?
5. Is this at the centre of the tube or closer to one or other of the chemicals used?
6. What can you say about the speed with which the gases ammonia and hydrogen chloride diffuse?

2.1

In Activity 2.1.1 each gas is in high concentration immediately around the cotton wool and in low concentration in the remainder of the tube. The particles of each gas move from the regions of high concentration to low concentration across a concentration gradient.

Eventually, as the particles of each gas diffuse, they meet and a chemical reaction takes place.

Diffusion into and out of cells

a)
diffusion of oxygen
high concentration of oxygen   low concentration of oxygen

b)
diffusion of carbon dioxide
low concentration of carbon dioxide   high concentration of carbon dioxide

FIG 2.1.2 Diffusion of gases a) into and b) out of cells

Substances diffuse into and out of cells across concentration gradients. For example:
- The concentration of oxygen in the blood is higher than in the cell, where it is used up during respiration, so oxygen diffuses from the blood into the cell.
- The concentration of carbon dioxide in the cell is higher, because it is produced during respiration, so carbon dioxide diffuses from the cell into the blood.

Fun fact

Diffusion takes place through solids, but this happens much more slowly than through liquids and gases.

Each spread offers questions to help you to check whether you have understood the topic.

Check your understanding

1. Bromine is an orange-brown volatile liquid. The following diagram shows what happened when a small amount of bromine was placed in a gas jar and another inverted gas jar was placed on top. The apparatus was left for 1 hour.
   a) Explain what has happened.
   b) Predict what the apparatus will look like after several hours.

FIG 2.1.3

Key terms

diffusion the spreading out of particles of a substance to fill the space available

concentration the number of particles of a substance per unit volume (for example, in a solution, g/cm³ of water)

concentration gradient the difference in the concentration of particles of a substance in one place compared to another place

Key terms are defined on the pages where they are used.

Review of Transport in humans

- The circulatory system consists of the heart and a network of blood vessels that carry blood to all the cells of the body.
- The heart is a muscular sac that contracts and relaxes throughout a person's life without ever stopping. It consists of four chambers: right auricle, right ventricle, left auricle, left ventricle.
- The diastole is the moment when the heart muscle relaxes. During this time:
  ○ deoxygenated blood from the body enters the right auricle
  ○ oxygenated blood from the lungs enters the left auricle.
- The systole is the moment when the heart muscle contracts. During this time:
  ○ the auricles contract, forcing blood into the ventricles
  ○ when full, the ventricles contract, forcing blood out of the heart.
- Blood is carried away from the heart in arteries and towards the heart in veins.
- Arteries have thick muscular walls and a small lumen. They must withstand very high pressures as blood is pumped into them from the heart.
- Veins have thinner walls and a larger lumen. The pressure in veins is less than in arteries. Long veins have valves to prevent blood flowing in the wrong direction.
- Arteries divide into arterioles and then into capillaries. The wall of a blood capillary is only one cell thick so substances are able to pass between capillaries and cells. Capillaries combine to form venules. Venules combine to form veins.
- Blood consists of about 90 per cent liquid, which is called plasma, and ten per cent solids, which is mostly blood cells.
- There are different types of blood cells:
  ○ red blood cells that carry oxygen around the body
  ○ white blood cells that engulf germs – phagocytes
  ○ white blood cells that release chemicals that kill germs – lymphocytes.
- A pulse is caused by blood being pumped through arteries by the heart.
- A pulse can be felt at different points of the body where an artery passes over a bone near the surface of the skin. One of the easiest places to feel a pulse is the inside of the wrist.
- Pulse rate is the number of pulses per minute. Pulse rate increases during exercise as the body needs more glucose and oxygen to provide energy.

At the end of each unit there are pages which list the key topics covered. These will be useful for revision.

At the end of each section there are special questions to help you and your teacher review your knowledge. See if you can apply this knowledge to new situations and if you can use the science skills that you have developed.

Science, Technology, Education, Arts and Mathematics (STEAM) activities are included, which present real-life problems to be investigated and resolved using your science and technology skills. These pages are called **Science in practice**.

Review questions on Transport in humans

1. Fig 3.RQ.1 shows a section through a blood vessel.

FIG 3.RQ.1

a) Does Fig 3.RQ.1 represent an artery, a vein or a blood capillary? Explain your answer.

b) In which direction does the blood flow through this vessel? Explain how you know.

2. a) Explain why the blood pumped from the heart to the body is brighter red than the blood that returns to the heart.

b) Fig 3.RQ.2 shows details of the human heart.

i) Identify the chambers of the heart marked W and X.

ii) State where the blood comes from to Y.

iii) State where the blood goes to from Z.

c) Some people are born with a hole between the two sides of their heart. Explain why this leaves them feeling weak and lacking in energy.

FIG 3.RQ.2

3. Jessica skipped for 15 minutes. When she stopped, her pulse rate was taken for ten minutes. The results are given in the table.

Pulse rate (beats per minute)	113	100	88	79	73	69	66	64	63	62	62
Time (minutes)	0	1	2	3	4	5	6	7	8	9	10

TABLE 3.RQ.1

a) Draw a graph of pulse rate on the vertical axis against time on the horizontal axis.

b) Explain why skipping increases the pulse rate.

c) What is Jessica's pulse rate at rest?

Making an artificial heart

Some people who have diseased hearts can get a transplant in which they receive a healthy heart from a person who has recently died from other causes, such as a road traffic accident. However, there is a risk that the donor heart may be rejected by the patient's body. Also, there are insufficient donor hearts to satisfy demand. An alternative to a heart transplant could be to replace a diseased heart with an artificial heart.

The Heart Foundation of Jamaica wishes to raise money for research into building an artificial heart. They have asked you to apply your knowledge of the heart to build a simple model which can be used at fund raising meetings. The model needs to demonstrate how a liquid can be pumped from one container into another, in a circuit, using a small electric motor.

1. You are going to work in groups of 3 or 4 to build a simple model that represents the action of the heart. The tasks are:

- To review the structure of the heart and the movement of the blood around the circulatory system.
- To design a model powered by an electric motor that pumps a liquid from one vessel into another, continuously in a circuit.
- To build your model.
- To test your model.
- To modify your model on the basis of test results.
- To demonstrate your model as part of a presentation in which you explain how you went about creating it.

a) Look back through the unit and make sure you understand how the heart functions in terms of pumping blood around the body.

b) In essence you are being asked to create a model of single circulation. This can be between the right ventricle and left auricle (via the lungs), or the left ventricle and the right auricle (via the body).

What are you going to use for your containers? They can be open, such as two beakers, or you might decide to use empty plastic bottles.

What are you going to use to carry the liquid? It might be better to use clear plastic tubing rather than rubber tubing so the liquid can be observed.

You can make some lifelike 'blood' with water and a few drops of red food colouring.

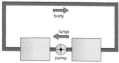

FIG 3.SIP.1 Single circulation

Unit 1: Working like a scientist

We are learning how to:

- identify and use units for different quantities
- use prefixes to alter the values of units.

SI base units

When scientists carry out experiments they often take measurements. These measurements are made in **SI units**. These are an internationally agreed set of units used by all scientists around the world. Using the same units makes it easier for scientists to communicate their work to others and carry out calculations that relate different quantities.

There are seven SI base units but only five are commonly used. These are given in Table 1.1.1.

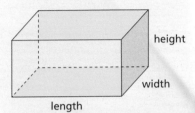

FIG 1.1.1 Volume of a cuboid

Unit name	Unit symbol	Quantity measured	Additional information
metre	m	length and distance	
kilogram	kg	mass	
second	s	time	
kelvin	K	temperature	degree Celsius (°C) normally used
mole	mol	amount of substance	this is not the same as mass

TABLE 1.1.1

You will already be familiar with the first four quantities in this table from everyday life. The mole is a measure of the number of particles taking part in chemical reactions and it is not the same as mass. You will learn more about the mole later in your course.

FIG 1.1.2 Fruits and vegetables are sold by the kilogram

SI derived units

In addition to the base units, there are a number of other units derived from them. For example, the unit of volume is derived from the unit of length.

The volume of a regular shape like a cuboid = length × width × height. Since each of those quantities is measured in a unit of length the volume will be given in that unit cubed. Since the metre is the SI unit of length, the cubic metre will be the SI unit of volume.

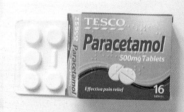

FIG 1.1.3 Amounts of drugs are often expressed in milligrams

FIG 1.1.4 A ruler shows the length in centimetres and millimetres

Although the SI unit of mass is the kilogram, this unit is not suitable for expressing the mass of all objects since the values would require large numbers of zeroes. For example:

The mass of the Earth = 5 972 000 000 000 000 000 000 000 kg

The mass of an atom of carbon = 0.000 000 000 000 000 000 000 019 9 kg

To get around this problem we use a series of **prefixes** which denote different powers of 10. The most commonly used prefixes are shown in Table 1.1.2.

Notice that the symbol for mega is 'M' while the symbol for milli is 'm'. It is important to write all of these prefixes in upper or lower case as appropriate to avoid confusion.

Prefix	Symbol	Multiplying factor	Multiplying factor in scientific notation
mega	M	1 000 000	10^6
kilo	k	1 000	10^3
deci	d	0.1	10^{-1}
centi	c	0.01	10^{-2}
milli	m	0.001	10^{-3}
micro	μ	0.000 001	10^{-6}

TABLE 1.1.2

All of these prefixes may be used with any SI base or derived unit; however, some are more commonly used than others.

Activity 1.1.1

SI units in everyday life

You may be asked to carry out this activity at home. Here is what you should do:

1. Look at the packaging on foods and other materials in your home. Make a note of any SI unit used to measure things.

2. Look in the area around your home at road signs, shops, garages etc. Make a note of any SI units you see being used to measure things.

3. Are SI units widely used? Record some examples.

4. Are there areas where units other than SI units are being used? Give examples.

FIG 1.1.5 Liquids are measured in cubic centimetres and cubic decimetres

Check your understanding

1. State the SI unit used to measure each of the following.

   a) Mass   **b)** Temperature   **c)** Volume

2. Copy and complete the following:

   a) 1 mg = ... g   **b)** 1 km = ... m

   c) 1 mm = ... m   **d)** 1 kg = ... g

   e) 1 cm = ... mm   **f)** $1 cm^3$ = ... dm^3

   g) 1 μg = ... mg   **h)** 1 cm = ... km

Key terms

SI units group of internationally agreed units used in science

prefix group of letters or short word put in front of other words to modify their meaning

Measuring quantities

We are learning how to:

• measure different physical quantities.

Measuring quantities ▶▶▶

A **physical quantity** is a quantity that can be measured. To obtain knowledge, scientists carry out experiments in which they make careful observations and measurements. In order to do this they use a variety of measuring instruments and containers which are collectively called apparatus.

FIG 1.2.1 Balances measure mass

Mass ▶▶▶

The **mass** of a substance is the amount of matter it contains. The units of mass are the kilogram (kg), for large masses, and the gram (g), for small masses.

$$1\,kg = 1\,000\,g$$

Mass is measured on a balance. There are two types in common use in laboratories.

A beam balance consists of a weigh pan on one side and a moveable mass on the other. The sample is placed on the weigh pan and the mass is moved along a scale until the beam remains horizontal. The mass of the sample is then read off the scale.

An electronic balance consists of a weigh pan mounted on a box containing electronic circuitry. The sample is placed on the weigh pan and its mass is given on a digital display.

FIG 1.2.2 Digital stopwatch

Volume ▶▶▶

The **volume** of a substance is the amount of space that it occupies. The units of volume are the cubic decimetre (dm^3), for large volumes, and the cubic centimetre (cm^3), for smaller volumes. For very large volumes the cubic metre (m^3) may also be used.

$$1\,m^3 = 1\,000\,dm^3$$

$$1\,dm^3 = 1\,000\,cm^3$$

Time ▶▶▶

For the purposes of most experiments, time is measured in minutes and seconds. A wristwatch will provide sufficiently accurate measurements for this.

Key terms

mass amount of matter a substance contains

volume amount of space a substance occupies

temperature hotness of a body

physical quantity quantity that can be measured

Temperature

Temperature is a measure of the degree of hotness of a body. Although the systematic unit of temperature is the Kelvin (K), temperature is generally measured in degrees Celsius (Centigrade) (°C). The relationship between degrees Celsius and degrees Kelvin is very simple.

Celsius = Kelvin − 273

Kelvin = Celsius + 273

The most common device for measuring temperature in the laboratory is the liquid-in-glass thermometer.

FIG 1.2.3 Liquid-in-glass thermometers

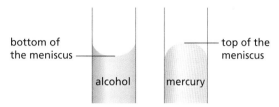

bottom of the meniscus

alcohol

top of the meniscus

mercury

FIG 1.2.4 Reading the level of liquid in a calibrated tube

Activity 1.2.1

Taking measurements

Here is what you should do:

1. Weigh a dry 100 cm³ beaker and record its mass.

2. Measure out approximately 25 cm³ of water and pour the water into the beaker. Reweigh the beaker and water and record the total mass.

3. Calculate the mass of the water used.

4. Measure the initial temperature of the water and record it.

5. Place the beaker on a tripod and gauze and heat it for exactly 5 minutes.

6. Measure the final temperature of the water. What was the temperature increase of the water?

Check your understanding

1. **a)** The mass of a beaker containing some powder is measured on an electronic balance (Fig 1.2.5). What is the reading on the balance?

   b) After removing the powder the mass of the empty beaker is 119.87 g. What is the mass of the powder?

2. **a)** What is the level of aqueous solution in the burette shown in Fig 1.2.6?

   b) 4.8 cm³ of the solution is run out of the burette. What is the reading on the burette now?

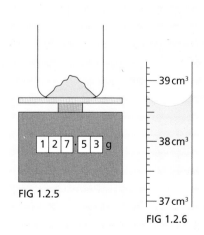

FIG 1.2.5

| 1 | 2 | 7 | · | 5 | 3 | g |

39 cm³

38 cm³

37 cm³

FIG 1.2.6

Data gathering

We are learning how to:

- make qualitative and quantitative observations
- collect data.

Data gathering >>>

Data is another term for information. When we carry out experiments we gather information, often by observing what happens. In everyday language 'observing' usually means what you see, but in science it can mean:

- what you see
- what you smell – some gases have characteristic smells
- what you hear – some solids crackle when heated; this is called decrepitation
- what you feel – sometimes test tubes get warm because chemical reactions give out heat.

Qualitative observations >>>

Many of the observations you will make don't involve numbers or measurement. These are called **qualitative** observations.

These include things like changes of colour, the release of bubbles of gas (called effervescence) and the emission of heat, and sometimes light as well.

Quantitative observations >>>

Some observations involve numbers and therefore measurement of some kind.

FIG 1.3.1 Qualitative changes

Key terms

data information

observation what you see, hear, smell or feel

qualitative to do with quality like colour

quantitative to do with measurement

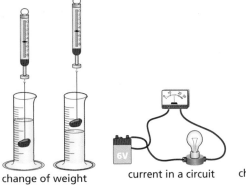

change of weight

current in a circuit

change of temperature

FIG 1.3.2 Quantitative changes

Quantitative changes include changes in weight, current in a circuit and changes in temperature. In some experiments you might measure the quantity at the start and again at the finish. This will allow you to calculate any increase or decrease.

In other experiments you might monitor the quantity continuously, recording readings at regular time intervals. If you do this, you must write down the readings in the form of a table.

Activity 1.3.1

Making observations

Here is what you need:

- Iron nail
- Dilute hydrochloric acid
- Dilute sodium hydroxide
- Calcium carbonate
- White bread
- Iodine solution
- Thermometer
- Copper powder
- Tin lid
- Balance
- Heat source
- Tripod
- Stand and clamp
- Eye protection.

Make observations on each of the following reactions and state whether the changes you observe are qualitative or quantitative.

1. Hold an iron nail in a clamp and heat one end of it strongly.
2. Place a few drops of iodine solution onto a piece of white bread
3. Place a small amount of calcium carbonate in a test tube and add dilute hydrochloric acid to a depth of about 2 cm.
4. Place dilute hydrochloric acid in a test tube to a depth of about 3 cm. Measure the temperature of the acid. Leave the thermometer in the test tube and add 3 cm of dilute sodium hydroxide. Measure the temperature of the reaction mixture.
5. Place a small amount of copper powder on a tin lid. Find the total mass of the tin lid and copper. Put the tin lid on a tripod and heat it for a few minutes. Allow the tin lid to cool. Find the total mass of the tin lid and product.
6. Record all your observations in a table.

Check your understanding

1. State whether each of the following observations is qualitative or quantitative.
   a) When a metal rod was heated its length increased by 0.7 mm.
   b) The mass of the reaction mixture decreased by 1.35 g.
   c) During exercise a person's heart rate increased from 75 to 102 beats per minute.
   d) The presence of simple sugars turns Benedict's reagent from blue to orange.
   e) The temperature of the reaction increased from 25.5 °C to 41.0 °C.

2. A chemical reaction was carried out in a beaker. The reactants were added at time = 0 seconds and then the mass of beaker and reaction mixture was measured every 15 seconds for 3 minutes. Present the mass data in a suitable table. The mass values were:
   149.3 g, 145.2 g, 141.3 g, 137.8 g, 134.7 g, 132.0 g, 129.8 g, 127.9 g, 125.9 g, 124.4 g, 123.3 g, 122.5 g, 122.3 g.

Plotting a graph

We are learning how to:

- plot a graph
- interpret a graph.

A graph is a method of displaying data but it can also provide important information about physical quantities related to those in the graph.

Distance–time graphs

Fig 1.4.1 shows how the distance of an object from its starting position changed over time. This is called a **distance–time graph**.

The line shows the distance travelled at any given time, but the slope or **gradient** of the line provides more information.

It doesn't matter which two points on the graph are used to measure the gradient. The result will always be the same.

The gradient of the graph above is:

$\frac{6 \text{ metres}}{3 \text{ seconds}}$ = 2 metres per second, m/s or ms⁻¹

In terms of quantities the gradient of the graph is $\frac{distance}{time}$, which is equal to speed. The gradient of a distance–time graph therefore tells us the speed of the object. In this example speed = 3 m/s.

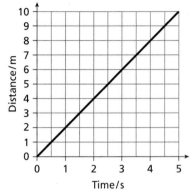

FIG 1.4.1 A distance–time graph

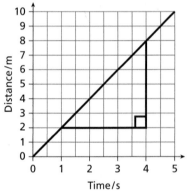

FIG 1.4.2 Measuring the gradient of a straight-line graph

Velocity–time graphs

You should recall from Grade 8 that the velocity of an object is its speed in a particular direction. A **velocity–time graph** also provides information about the motion of an object.

The graph itself shows the velocity of the object at any given time.

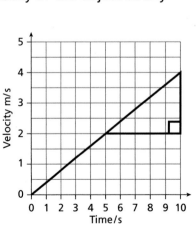
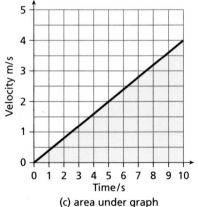

(a) velocity–time graph   (b) gradient of graph   (c) area under graph

FIG 1.4.3 Information from a velocity–time graph

The gradient of a velocity–time graph is equal to $\frac{\text{velocity}}{\text{time}}$, which is **acceleration**.

Acceleration is the rate of change of velocity with respect to time. In this example:

Acceleration = $\frac{2 \text{ metres/second}}{5 \text{ seconds}}$

= 0.4 meters per second per second, m/s^2 or ms^{-2}

The area under a velocity–time graph is also important since velocity × time = displacement.

Activity 1.4.1

To investigate the relationship between distance and time

Here is what you need:

- Markers such as coloured balls, cloths etc. × 5
- Measuring tape
- Stopwatch.

1. Place a marker where you are standing and start walking in a straight line at a steady pace. Start the stopwatch.

Time / s	Distance walked / m
0	0
5	
10	
15	
20	
25	

TABLE 1.4.1

2. Drop a marker at your position after walking for 5 seconds. Repeat this every five seconds until you have used up all your markers.

3. Measure the distance you have walked to each marker from where you started. Copy Table 1.4.1 and enter your data.

5. Plot a distance–time graph for your walk. Draw a straight line of best fit through the points.

6. Use your graph to calculate your walking speed.

Check your understanding

1. Table 1.4.2 shows how the velocity of an object changed over time.

Velocity m/s	Time /s
0	0
0.5	1
1.0	2
1.5	3
2.0	4
2.5	5

TABLE 1.4.2

   a) Plot a graph of velocity against time.

   b) Use your graph to find:

      i) The displacement of the object after 5 seconds.

      ii) The acceleration of the object.

Key terms

distance–time graph a graph of distance on the y-axis and time on the x-axis

gradient slope of a graph

velocity–time graph a graph of velocity on the y-axis and time on the x-axis

acceleration rate of change of velocity with respect to time

Structuring an experiment

We are learning how to:

- identify the different stages in the structure of an experiment
- test a hypothesis.

An experiment requires careful planning if it is to provide the scientist with useful or valid information. There are several stages to consider.

- Specify the problem: What is the purpose of the experiment? What is the problem you are hoping to solve?

The experiment that you will design must have the following:

- A **hypothesis** – this must clearly relate to the observation that was made. What do you think the answer to the problem might be? It doesn't matter if your idea is not correct. Experiments that produce negative results are often as important as those that produce positive results.

- An Aim – this must clearly relate to the hypothesis.

- A list of the apparatus and materials that you will use in carrying out your experiment. Try to avoid using words such as some, a few and many. Be precise and instead use 500 g of soil, 200 cm³ of water, 10 millipedes etc.

- A clear Method. Unlike the other experiments that you have written, Planning and Design experiments are written in the present tense and not the past tense. This is because the experiment has not yet been done, so you are giving instructions in your method. How are you going to carry out your experiment? Are there any safety issues that must be considered?

- You experiment must be a **fair test**. You must keep all the variables constant, except for the one you are intentionally changing — the **independent variable**. Any other **dependent variables** must be changing because of the changes in the independent variable.

- Ensure that you are clear which variables are to be kept constant (**control variables**), which you plan to change and which you expect to depend on these changes.

- A summary of the expected results. Will you record your data in a table? How will you present your data? This can be written in point form and would indicate whether your hypothesis is proven or not proven.

> **Fun fact**
>
> Sometimes two similar experiments are conducted at the same time. In the second experiment there are no variables. This is called a control experiment and is designed to find out if changing a variable in the first experiment is really responsible for any change observed.

Key terms

hypothesis proposed explanation

fair test test that gives valid results

independent variable quantity allowed to vary

dependent variable quantity that varies because of changes to the independent variable

control variable quantity which is kept constant

template pattern or way of organising

- A list of the limitations of your experiment.

The above steps provide a **template** that can be applied to most experiments.

Activity 1.5.1

Testing a hypothesis

Eating too much food sometimes results in the formation of excess acid in the stomach, which we call indigestion. This can be relieved by taking antacid tablets, which contain mild alkalis.

A student wants to know which of three types of antacid, types X, Y and Z, is the most effective. She will do this by finding which will neutralise the most acid. Her initial hypothesis is that they are all equally effective.

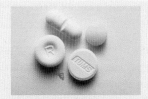

FIG 1.5.1 Antacids are used to treat stomach ache

Your task is to design and carry out an experiment to test the hypothesis made by the student. Here is what will need:

- Beakers 250 cm³ × 3
- Dilute hydrochloric acid
- Acid-base indicator
- Mortar and pestle
- Samples of antacids X, Y and Z
- Distilled water
- Balance
- Burette.

1. Crush each type of antacid to form powders.
2. Weigh exactly 1 g of each powder and place it in separate marked beakers.
3. Dissolve each of the powders in 100 g of water.
4. Add a few drops of acid-base indicator to each solution. This will change colour when the solution becomes acidic.
5. Add dilute hydrochloric acid of the same concentration and the same temperature from a burette to each solution in turn until the solution becomes acidic.
6. Record the volume of acid used each time. Present your data as a table.
7. Use your data to determine whether the initial hypothesis was correct or not.

FIG 1.5.2

Check your understanding

1. In the above activity:
   a) What was the independent variable?
   b) What was the dependent variable?
   c) What were the control variables?

Designing an experiment

We are learning how to:

- structure an experiment to get meaningful results.

Charcoal pieces provide the heat needed to cook food on a barbeque. The charcoal may be made from different woods.

It is difficult to measure the amount of heat energy given out by a fuel directly. An **indirect** method is to measure the temperature increase when the heat energy is used to raise the temperature of water in a suitable container such as a boiling tube or a beaker.

It takes about 4.2 J of heat energy to raise the temperature of 1 g of water by 1 °C. So, for example, the heat needed to raise the temperature of 50 g of water from 25 °C to 40 °C would be:

$$4.2 \times 50 \times (40 - 25) = 4.2 \times 50 \times 15 = 3150 \text{ J}$$

FIG 1.6.1 Barbeque charcoal

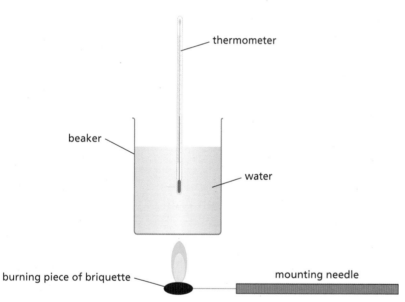

FIG 1.6.2 Apparatus to measure heat energy indirectly

Activity 1.6.1

Measuring the energy given out by different types of charcoal

In this activity you are going to design an experiment to determine which of three types of charcoal gives out the most heat. Use the notes under the headings to help you.

1. **Specify the problem:** What is the purpose of this experiment?

Fun fact

The type of charcoal is a qualitative variable while the rise in the temperature is a quantitative variable.

2. **Formulate a hypothesis:** Is there anything that might suggest one type of charcoal is better than the others? Your hypothesis might be that there is no difference between different types.

3. **State the Aim and Method:**
   - How are you going to measure the amount of heat given out by each type of charcoal?
   - What is going to be your independent variable?
   - What is going to be your dependent variable?
   - What other variables will you control to make sure it is a fair test?

4. **Record and analyse the results:**
   - How are you going to record data from your experiment?
   - Is there some way of presenting the data that will make it clear

5. **Draw conclusions:** How will you decide if the results of the experiment support your hypothesis or not?

Check your understanding

1. A student wishes to compare the reactivity of three metals – zinc, iron and magnesium – with acids by measuring the amount of hydrogen given off over time. Here is what he plans to do.

   React zinc powder with cold dilute sulfuric acid in a boiling tube and measure the amount of hydrogen produced in 5 minutes.

   React a lump of iron with hot dilute sulfuric acid in a conical flask with stirring and measure the amount of hydrogen gas produced in 3 minutes.

   React magnesium ribbon with cold concentrated hydrochloric acid in a round-bottomed flask and measure the amount of hydrogen gas produced in 2 minutes.

   a) Explain why the student is not carrying out a fair test.

   b) Rewrite his plan so that it would provide reliable results.

Key term

indirect not directly caused by or resulting from

Reversible changes and irreversible changes

We are learning how to:

- understand the differences between physical and chemical changes
- differentiate reversible changes from irreversible changes

The changes that you observe when carrying out experiments can be described as **physical changes** or **chemical changes**. In this lesson you will explore whether physical changes and chemical changes are reversible or irreversible.

Physical changes

When water is placed into a freezer it turns to ice. When water is heated in a pan it eventually boils and turns into steam. Does this mean that ice, water and steam are three different substances or that they are simply three different forms or states of the same substance?

Ice, liquid water and steam all consist of the same water particles but arranged and moving about in different ways. They are therefore three states of the same substance.

If you place ice cubes in a glass and leave them at room-temperature they soon become liquid water. Similarly steam rapidly condenses on a cool surface, such as a kitchen window, to become liquid water.

Physical changes like changes of state:

- don't produce energy;
- don't produce any new substances;
- are easy to reverse.

FIG 1.7.1 Physical changes

FIG 1.7.2 Chemical changes

Chemical changes

Burning or combustion, and rusting are two changes we often observe in our everyday lives. How do these changes differ from the physical changes previously described?

When a fuel like wood is burnt, it produces energy in the form of heat and light. The wood is changed into ash and gases are released into the air. The particles that form the ash and the waste gases are different from those that formed the wood. It is impossible to turn ash and gases back into wood.

When iron turns into rust the change is very slow. Heat energy is produced but not quickly enough to cause any detectable temperature change. The smooth shiny grey iron becomes covered in a dull, rough brown layer of rust. The particles in rust are different from those in iron. It is very difficult to turn rust back into iron.

Chemical changes like burning and rusting:

- produce energy
- produce new substances
- are difficult or impossible to reverse

The differences between physical changes and chemical changes are summarised in Table 1.7.1.

Physical changes	Chemical changes
Physical changes are generally easy to reverse	Chemical changes are difficult or impossible to reverse
Physical changes don't produce any new substances	Chemical changes produce new substances
Physical changes don't produce energy	Chemical changes often produce energy in the form of heat, light and sometimes sound

TABLE 1.7.1

Activity 1.7.1

Reversible and irreversible changes

Here is what you will need:

- Tripod and gauze
- Tin lid (x2)
- Heat source
- Tongs
- Sand
- Saw dust
- Candle wax
- Aluminium foil
- Copper powder
- Sugar
- Copper carbonate

Here is what you should do:

1. Place a small sample of each material on a tin lid and heat it gently for about two minutes.
2. Observe any changes that take place during heating.
3. Allow each substance to cool and then observe again.
4. State whether heating each substance produced a reversible or an irreversible change. Give your results in the form of a table like the one below.

Substance	Reversible or irreversible change	Reason

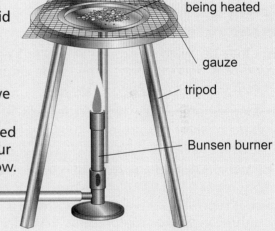

substance being heated

gauze

tripod

Bunsen burner

FIG 1.7.3

Check your understanding

1 Use the features given in Table 1.7.1 to determine whether each of the following is a reversible or an irreversible change.

a) Coconut oil becoming solid when placed in the refrigerator

b) Striking a match

c) Butter melting on a warm day

d) Boiling an egg

e) Paint drying

f) Magnetising an iron bar

Key terms

reversible change a change where a substance can easily be converted back into its original form

irreversible change a change where a substance cannot easily be converted back into its original form

Accuracy and scientific notation

We are learning how to:

- be aware of the level of accuracy possible with different pieces of apparatus
- express values in scientific notation.

Although accurate measurement is an essential part of making quantitative observations, we are often limited by the accuracy of the equipment we use.

A metre rule is ideal for measuring to the nearest centimetre or even millimetre. If we need to measure more accurately we can use an instrument like Vernier callipers, which will measure length to the nearest one tenth of a millimetre or better.

When taking readings of any kind you should be mindful of the limitations of the equipment you are using. It would make no sense to write down values to a greater degree of accuracy than the equipment is capable of providing.

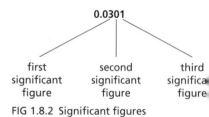

FIG 1.8.1 Measuring length

Significant figures 》》》

You may be asked to express values to a certain number of significant figures. This is a means of simplifying an answer by reducing the number of figures.

In a large number like 123456 the digit '1' is the most significant since it represents 100 000. The digit '2' is next in significance since it represents 20 000, and so on.

The same is true of a very small number. For example, in the number 0.000 987 654 the digit '9' is the most significant since it represents 9 ten thousandths. The digit '8' is next in significance since it represents 8 hundred thousandths and so on.

0.0301

first significant figure — second significant figure — third significant figure

FIG 1.8.2 Significant figures

In any number therefore, the most significant digit is the one that appears on the left, provided it is not zero. Any zeroes to the right of this will be significant, however.

You might be asked to round off values to any number of significant figures but rounding to 1, 2 or 3 significant figures is common. The term 'significant figures' is often shortened to 'sig figs' or 's.f.'.

The rules for rounding to the nth significant figure are:

- If the (n+1)th significant figure is 5 or more you round up the nth figure.
- If the (n+1)th significant figure is less than 5 leave the nth figure unchanged.

Let's look at an example.

The mass of a block is 129.65 g. Give this value to 3 sig fig.

The 4th significant figure is 6 so we round up the 3rd sig fig.

The mass of the block is 130 g (correct to 3 s.f.).

In Table 1.1.2 as well as listing prefixes for use with SI units, the multiplying factors were also given in **scientific notation**, which is also called **standard form**.

Scientific notation is a format for writing numbers in which there is a single digit between 1 and 9 to the left of the decimal point, numbers to the right of the decimal point, and multiplication by a power of 10. For example, the number of atoms in 12 g of carbon is 6.02×10^{23}.

If there are no digits to the right of the decimal point, a single zero is written. For example, the speed of light in a vacuum is approximately 3.0×10^8 m/s.

Scientific notation is also used to express very small values. The only difference is that the powers of 10 are negative. For example, the diameter of a carbon atom is 1.54×10^{-10}, which is the same as $1.54 \times \frac{1}{10^{10}}$.

> **Fun fact**
>
> Any number can be expressed in scientific notation:
>
> - Write the number with a single digit to the left of the decimal point.
>
> - Multiply by 10 to the power needed to generate the number when expanded.

Activity 1.8.1

Testing the accuracy of a beaker and measuring cylinder

Here is what you need:

- Beaker 100 cm³ calibrated in 50 cm³
- Measuring cylinder 100 cm³.

1. Weigh the beaker dry and empty.

2. Use the beaker to measure out 50 cm³ of water as accurately as you can. Reweigh the beaker and water and calculate the mass of the water.

3. Repeat these steps using the measuring cylinder.

4. The mass of 50 cm³ is 50.00 g. Which piece of apparatus measured the water:

   a) most accurately?   **b)** least accurately?

Check your understanding

1. Express each of the following to the stated number of significant figures.

   a) 21.43 cm³ (1 s.f.)   **b)** 0.837 g (2 s.f.)

   c) 129 s (2 s.f.)   **d)** 31.5 °C (2 s.f.)

   e) 1.919 g (2 s.f.)   **f)** 414.55 cm³ (2 s.f.)

2. Rewrite the following in scientific notation.

   a) 29.6 g   **b)** 0.86 cm³   **c)** 129 s

   d) 0.0036 g   **e)** 814 °C   **f)** 10.05 cm³

> **Key terms**
>
> **scientific notation** method of writing numbers as a decimal between 1 and 10 multiplied by a power of 10
>
> **standard form** same as scientific notation

Review of Working like a scientist

- Scientists express quantities in SI units.

- There are a small number of base units and many other units derived from them.

- Prefixes are used in front of units to denote different powers of ten.

- The prefixes most commonly used are mega, kilo, deci, centi, milli and micro.

- Many everyday substances are measured in SI units.

- Mass is the amount of matter a substance contains and is measured by a balance.

- Volume is the amount of space a substance occupies and can be measured by a variety of apparatus, including measuring cylinders and burettes.

- Time is measured in seconds, or sometimes minutes, using a digital stopwatch.

- Temperature is measured in Kelvins or degrees Celsius using a thermometer.

- Data is another word for information.

- Qualitative observations relate to things like changes of colour, shape or the emission of bubbles of gas but do not involve numbers.

- Quantitative observations relate to things like changes in mass, length, potential difference or temperature and involve numbers.

- In science quantitative data is sometimes displayed in the form of a graph.

- A graph shows how the value of one variable changes as the value of another is varied.

- The gradient of a graph is its slope.

- On a distance–time graph the gradient gives the speed of the moving object.

- On a velocity–time graph the gradient gives the acceleration of the moving object and the area under the graph gives the total displacement.

- Carrying out an experiment involves the following stages:
  - Specifying the problem
  - Formulating a hypothesis
  - Designing the experiment
  - Analysing data
  - Drawing conclusions.

- Values are sometimes expressed to a certain number of significant figures.

- Values may be expressed in scientific notation (which is also called standard form).

Review questions on Working like a scientist

1. State the SI unit used to measure each of the following.

   a) Length   b) Time   c) Mass

2. Express each of the following in the unit indicated.

   a) 12 mm in cm   b) 5 kg in g   c) 100 mg in g

   d) 50 cm³ in dm³   e) 150 µg in mg   f) 25 cm in m

3. a) Express the following temperatures in Kelvins.

      i) 0 °C   ii) 85 °C

   b) Express the following temperatures in degrees Celsius.

      i) 373 K   ii) 450 K

4. The graph in Fig 1.RQ.1 shows how the velocity of an object changed over time as it travelled from point A to point B.

   a) What was the acceleration of the object during the first 6 seconds of motion?

   b) After how many seconds did the object start to slow down?

   c) What was the acceleration of the object during the last 6 seconds of motion?

   d) What was the total displacement of the object?

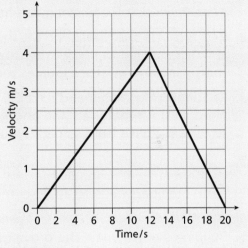

FIG 1.RQ.1

5. Briefly explain the significance of each of the following stages in planning an experiment.

   a) Specify the problem

   b) Formulate a hypothesis

   c) Designing the experiment

   d) Analysing data

   e) Drawing conclusions

6. Express each of the following to the stated number of significant figures.

   a) 37.21 dm³ (1 s.f.)   b) 0.0846 g (2 s.f.)   c) 12.19 s (3 s.f.)

7. Rewrite the following in scientific notation.

   a) 187.06 g   b) 0.016 s   c) 2390 cm³

What are STEAM activities?

STEAM stands for **S**cience, **T**echnology, **E**ngineering, **A**rt and Mathematics. The purpose of STEAM activities is to demonstrate how the science that you learn in the classroom can be applied to solve problems in many different fields.

At the end of each unit of this series you will find one or more STEAM activities designed to provide opportunities for students to apply the different subject knowledge. For each activity you will work in a group of 3 or 4 students and are expected to make a significant contribution towards the team effort.

There are several stages to every STEAM activity. The exact details will be different for each specific task, but we can summarise the stages in general terms.

1. **Defining the problem** – This tells you exactly what is required of the group. It will often be a problem that needs solving. It may involve making something or carrying out tests of some kind and observing results.

2. **Research** – In the remaining STEAM activities at the ends of the units in this book you will be expected to review the subject information given in the unit and supplement this by looking at other sources such as reference books and the Internet.

3. **Requirements** – You need to think about what you will need and make lists. You may need a list for materials and another list for tools and equipment. You might also need to obtain advice from someone outside your group, such as your teacher.

4. **Agree on a solution** – You are working as one student in a team of 3 or 4. The members of the team will probably suggest more than one solution to the problem. If so, then team members need to analyse and evaluate the suggestions of the different members and come to an agreed solution.

5. **Building a prototype** – This provides the team members with an opportunity to show their manual skills. The task should be organised in such a way that each member of the team makes a contribution. It may be useful if one person takes on the role of team coordinator.

6. **Testing** – Once the prototype is built, it needs to be tested or evaluated. It is often useful to have people from outside the team, who have not been closely involved in the planning and building, to review your work. A fresh pair of eyes can sometimes pick up things missed by people who are closely involved. Consider the comments made by your reviewers and modify what you have made in light of them, if you think it is appropriate.

7. **Reporting back** – The final task is to compile and deliver a report to explain what you did and to show your end product. This can take various forms, such as a PowerPoint presentation or a demonstration of what you have designed and built. If possible, take pictures at different stages of construction and testing on your cell phone. Illustrated reports are always more interesting than just writing and description.

FIG 1.SIP.1 Reporting the work of the team

1. When an image falls on the retina in the eye it persists for about 1/10th of a second. If another image falls on the retina before the first one fades then we see continuous movement rather than a sequence of still pictures.

A film reel consists of many still pictures or frames. When the reel passes through a projector the frames change so quickly that we see a moving picture or 'movie'.

A flick book contains a sequence of pictures in which each picture is slightly different to the previous one. Seeing the pictures quickly, one after another, gives the illusion of a moving picture. Flick books were a common amusement for children in the nineteenth century.

Your task is to produce a flick book to demonstrate movement in science.

2. There are lots of things you might choose to demonstrate. Here are some examples:

   - The movement of an animal such as a frog or a snake
   - The launching of a space rocket
   - The oscillation of a weight suspended on a spring
   - A germinating seed

3. In planning you might consider things such as:

   - How many images i.e. pages will we need?
   - How big will the pages need to be?
   - How can we ensure the image is always in the same position on the page?
   - How will we bind the pages to make the flick book?

FIG 1.SIP.2 A reel of film

4. As a team you must decide what your flick book will demonstrate.

5. The minimum you are going to need is paper, scissors, a pencil, a clip or adhesive, and some means of colouring. A single sheet of A4 paper can easily be folded to give 16 equal-sized pages. Will that be sufficient?

You might also decide you need some card from which you can cut templates. If the drawing is to be done by the whole team it is important that the subject is the same size and shape in each drawing.

FIG 1.SIP.3 A flick book

You might find it useful to number your pages so you can be sure they are in the correct order. This is particularly important when different people are working on different pages.

You will need something to bind your pages. A fold back clip works well. You can remove the metal arms so they don't get in the way.

6. Swap flick books with other teams and ask them to test your flick book and assess how well it demonstrates movement while you do the same for them. As a result of testing you might want to make some changes, such as replacing one or two pages, or perhaps making more use of colour.

7. Your report should state the movement you set out to create and how well you were able to do this. Your flick book should be passed around the class for other students to try. You should also mention any particular difficulties you experienced in making your book.

Unit 2: Transport across cells

We are learning how to:

- explain the movement of particles during diffusion
- relate diffusion to concentration gradient.

Diffusion ›››

You already know something about the process of **diffusion**, although you may never have heard it called by this name. Living things need to move substances around. One way in which this is done is by diffusion. Sometimes when you step into your home, you know that someone has been baking or cooking.

When someone is baking or cooking, tiny particles of food move through the air. The particles travel in the air from the kitchen to all the other rooms in your home. We call this spreading out of particles 'diffusion'.

FIG 2.1.1 Jerk chicken

Activity 2.1.1

Investigating diffusion

When ammonia gas and hydrogen chloride gas react they form a fine white 'smoke' of ammonium chloride particles. Your teacher will show you.

Here is what you need:

- Long glass tube with bungs at either end
- Cotton wool
- Concentrated aqueous ammonia
- Concentrated hydrochloric acid
- Stand and clamp.

Here is what you should do:

1. Support the tube horizontally using a stand and clamp.

2. To produce hydrogen chloride gas, wet a plug of cotton wool with hydrochloric acid, place it at one end of the tube and seal it with a bung.

3. To produce ammonia gas, wet a plug of cotton wool with aqueous ammonia, place it at the opposite end of the tube and seal it with a bung.

4. What do you observe inside the tube after a short time?

5. Is this at the centre of the tube or closer to one or other of the chemicals used?

6. What can you say about the speed with which the gases ammonia and hydrogen chloride diffuse?

In Activity 2.1.1 each gas is in high concentration immediately around the cotton wool and in low concentration in the remainder of the tube. The particles of each gas move from the regions of high concentration to low concentration across a concentration gradient.

Eventually, as the particles of each gas diffuse, they meet and a chemical reaction takes place.

Diffusion into and out of cells

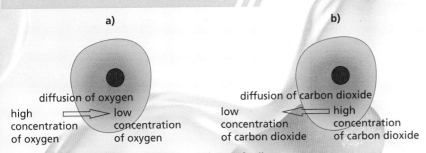

a)

diffusion of oxygen

high
concentration
of oxygen

low
concentration
of oxygen

b)

diffusion of carbon dioxide

low
concentration
of carbon dioxide

high
concentration
of carbon dioxide

FIG 2.1.2 Diffusion of gases a) into and b) out of cells

Substances diffuse into and out of cells across concentration gradients. For example:

- The concentration of oxygen in the blood is higher than in the cell, where it is used up during respiration, so oxygen diffuses from the blood into the cell.

- The concentration of carbon dioxide in the cell is higher, because it is produced during respiration, so carbon dioxide diffuses from the cell into the blood.

Fun fact

Diffusion takes place through solids, but this happens much more slowly than through liquids and gases.

Key terms

diffusion the spreading out of particles of a substance to fill the space available

concentration the number of particles of a substance per unit volume (for example, in a solution, g/cm³ of water)

concentration gradient the difference in the concentration of particles of a substance in one place compared to another place

Check your understanding

1. Bromine is an orange-brown volatile liquid. The following diagram shows what happened when a small amount of bromine was placed in a gas jar and another inverted gas jar was placed on top. The apparatus was left for 1 hour.

   a) Explain what has happened.

   b) Predict what the apparatus will look like after several hours.

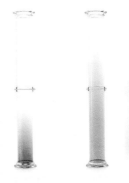

FIG 2.1.3

Osmosis

We are learning how to:

- explain the movement of particles during osmosis
- describe the action of a differentially permeable membrane.

Osmosis ▶▶▶

Osmosis is a special kind of diffusion that is concerned only with the movement of water molecules. Like other substances, water diffuses along a concentration gradient from a place where water is in high concentration to a place where it is in low concentration.

This can sometimes be a little confusing, because water is in a higher concentration in a dilute solution and in a lower concentration in a concentrated solution. What we are saying, therefore, is that under suitable conditions, water will move from a dilute solution, making it more concentrated, to a concentrated solution, making it more dilute.

A **differentially permeable membrane** is a membrane with holes that are large enough to allow small molecules like water to pass through, but small enough to prevent the movement of large molecules.

Fig 2.2.1 shows what happens if we separate pure water from a solution of a compound by a differentially permeable membrane. Notice that the water molecules move in both directions through the differentially permeable membrane. However, more water molecules pass from the dilute solution to the concentrated solution than pass in the other direction. Eventually the concentration of the solution reaches an **equilibrium** situation and remains unchanged.

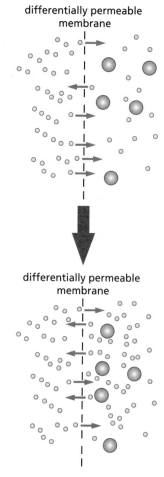

FIG 2.2.1 Movement of particles across a differentially permeable membrane

Activity 2.2.1

Investigating osmosis

Here is what you need:

- Shallow dishes x 2
- Raisins
- Distilled water
- Strong sugar solution.

Here is what you should do:

1. Half fill one shallow dish with distilled water and another with strong sugar solution.

2. Examine some raisins and then place a small handful in each dish (Fig 2.2.2).

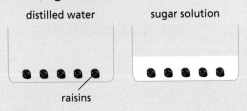

FIG 2.2.2

3. Leave the dishes somewhere warm for 30 minutes.

4. After 30 minutes examine the raisins in each dish. Are they still the same shape? Record your observations.

5. Leave the dishes overnight and examine the raisins again. Are they the same shape? Record your observations.

6. Explain your observations in terms of osmosis.

Key terms

osmosis diffusion that involves water molecules

differentially permeable membrane a membrane that allows some particles to pass through it but not others

equilibrium a situation in which the movement of particles in one direction is equal to the movement of particles in the opposite direction

Water passes into and out of all living cells by osmosis. The process of osmosis controls the concentration of substances in the cell.

Check your understanding

1. The membrane around a cell is a differentially permeable membrane. Fig 2.2.3 shows an experiment carried out by a student using two slices of cucumber. Each slice was weighed and then one slice was placed in distilled water and the other in strong salt solution.

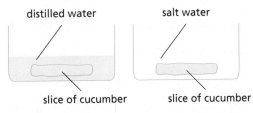

FIG 2.2.3

a) Predict how the masses of the slices of cucumber will change if they are left to stand for 1 hour.

b) Explain your answer.

c) Predict what would happen if the experiment was repeated using salt solution of exactly the same concentration as the liquid inside the cells of the cucumber.

Planning an experiment on osmosis

We are learning how to:

- identify the different stages in planning an experiment
- plan an experiment.

Planning an experiment on osmosis »»

An egg is covered by a hard shell composed of calcium carbonate. Immediately under the shell there is a thin membrane which holds the contents of the egg. If an egg is placed in weak acid for a few days, the shell reacts with the acid and is deshelled while the membrane remains undamaged.

The membrane surrounding the egg will allow water to pass into and out of the egg. Your teacher will provide you with two deshelled eggs, pure water and a strong sugar solution. You need to design an experiment in which you use these materials to investigate osmosis. Here are the different stages you should follow.

FIG 2.3.1 A partly deshelled egg

1. Specify the problem: What is the purpose of the experiment? What do you hope to find out about osmosis?

2. Formulate a **hypothesis**: Apply what you already know about osmosis and predict what you think the outcome of your experiment might be.

3. Designing the experiment: How are you going to carry out your experiment? What materials and equipment will you need in addition to the deshelled eggs, the water and the sugar solution?

   What will you do? You might decide to put the deshelled eggs in the water and the sugar solution. If you did, how long do you think it would take to see any changes that might take place?

   How can you be sure you are carrying out a **fair test**? In any experiment it is normal to allow one variable to change whilst keeping everything else constant. In this way you can be sure that changing that variable is responsible for any changes you observe.

   Are there any safety issues that you must take into account when designing your experiment?

FIG 2.3.2 Deshelled eggs in sugar solution and distilled water

4. Analysing data: What changes do you think might take place? For example, do you think that the masses of the

eggs might change? Do you think that the circumferences of the eggs might change? Do you think the volume of the eggs might change? Are there other things that might change? How can these changes be measured?

How will you record any data that you collect? Do you need to construct a table? In what units will you make your measurements? How accurate can you hope to be?

Can you think of an interesting way to present your data?

5. Drawing conclusions: What conclusions are you hoping to draw from this experiment? How do you think the data you gather will support your conclusions?

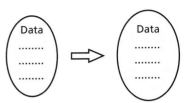

FIG 2.3.3 An egg before and after

Check your understanding

1. In an experiment on osmosis 30 chips of similar size were cut from one potato. There were divided into six batches of five chips and each batch was weighed.

   Each batch was placed in a sugar solution of different concentration and left for five hours. At the end of this time, each batch was removed, wiped and reweighed. The results of the experiment are given in Table 2.3.1.

   a) i) Copy and complete the table by calculating the missing changes in mass.

   ii) What substance is being gained or lost by the potato chips?

   iii) Predict what would happen if another batch of potato chips had been placed in a sugar solution that was even less concentrated than Batch A.

Batch	Sugar concen-tration	Mass of chips at start/g	Mass of chips at end/g	Mass gain (+) or loss (–)/g
A	Low	51.3	52.6	+1.3
B		49.8	50.7	
C		50.2	50.7	
D		51.6	51.4	
E		48.9	48.3	
F	High	50.7	49.7	–1.0

TABLE 2.3.1

   b) In which batch was the concentration of the sugar solution closest to that of the concentration of the solution in the cells of the potato chip? Explain your answer.

   c) Explain why:

   i) batches of five chips, not single chips, were used

   ii) batches of five small chips and not one large chip were used

   iii) the experiment was left for five hours and not five minutes.

Key terms

hypothesis proposed explanation

fair test test that gives valid results

31

Comparing diffusion and osmosis

We are learning how to:

• identify similarities and differences between diffusion and osmosis.

Comparing diffusion and osmosis ≫

Diffusion is the spontaneous movement of particles from an area of high concentration to one of lower concentration.

When you make a cup of tea the coloured substances contained in the tea leaves diffuse out into the hot water. We can speed up the process by stirring the water.

Diffusion is the result of the random motion of particles. As the temperature increases particles move more quickly. Diffusion therefore takes place more rapidly at higher temperatures.

Osmosis is the spontaneous net movement of water molecules across a differentially permeable membrane.

When plant or animal cells are placed in a solution, water molecules continually move into and out of the cells (Fig 2.4.2). Eventually the concentration of the solution inside the cell is equal to the concentration of the solution outside.

When the concentration of the solutions on either side of a membrane is the same (Fig 2.4.3), water molecules continue to move in each direction through the membrane. However, they move at the same rate so the concentrations remain unchanged. This is called **dynamic equilibrium**.

Diffusion and osmosis look very similar in that they are both means by which substances enter and leave cells; however, there are some important differences between these processes, which are summarised in Table 2.4.1.

FIG 2.4.1 Coloured substances in tea diffuse out into hot water

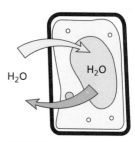

FIG 2.4.2 Water moves into and out of cells

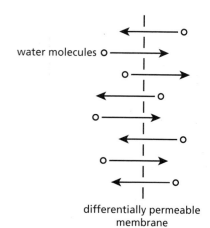

water molecules

differentially permeable membrane

FIG 2.4.3 Equal numbers of water molecules pass each way across the membrane

	Diffusion	Osmosis
When does it occur?	Diffusion occurs in the gaseous state or with particles in solutions when materials spread from an area of higher concentration to one of lower concentration.	Osmosis occurs when there is a difference in the concentrations of solutions either side of a differentially permeable membrane and water moves to equalise the concentrations.
Why is it important?	Diffusion is associated with the uptake of oxygen and the release of carbon dioxide by cells.	In animals osmosis allows the distribution of nutrients. In plants osmosis is responsible for the absorption and movement of water around the plant.
Concentration gradient	Diffusion takes place when materials move from a region of high concentration to a region of low concentration.	Osmosis takes place when water moves from a less concentrated solution to a more concentrated solution.
Water	Diffusion can take place in the absence of water.	Osmosis can only take place where there is water.

TABLE 2.4.1

Fun fact

Tea is correctly described as an infusion. This is an extract obtained by soaking leaves or whole plants in water.

Fun fact

Diffusion takes place quickly through gases and less quickly through liquids. Diffusion also takes place through solids, but this process is very slow.

Check your understanding

1. Decide whether the following is the result of diffusion or of osmosis.

   a) In the lungs oxygen gas passes from the air into the bloodstream.

   b) Water passes into red blood cells.

   c) Carbon dioxide passes from air into plant cells.

   d) When a raw potato chip is placed in strong sugar solution it shrinks.

   e) It is possible to smell petrol vapour on a garage forecourt even when you are not stood next to the pumps.

   f) When one peg of grapefruit is placed in sugar in a sealed bag the sugar forms a syrup after a short time.

Key terms

dynamic equilibrium system that is continually changing but remains in balance

Review of Transport across cells

- Diffusion is the movement of particles from a region of higher concentration to a region of lower concentration along a concentration gradient.

- Substances pass into and out of plant cells by diffusion.

- Osmosis is a special kind of diffusion which involves the movement of water particles.

- During osmosis water molecules move from a more dilute solution to a more concentrated solution through a differentially permeable membrane.

- Water passes into and out of plant cells by osmosis.

- Living tissue has differentially permeable membranes across which osmosis can take place.

- Diffusion involves gases and particles in solution.

- The rate of diffusion increases with temperature.

- Osmosis involves water in aqueous solutions.

- Diffusion is involved in the energy-storing and -releasing processes of photosynthesis and respiration.

- In plants osmosis is responsible for the absorption and movement of water.

- In animals osmosis is involved in the transport of nutrients into tissues and cells.

- Diffusion takes place from an area of high concentration to an area of low concentration.

- Osmosis takes place from a dilute solution to a more concentrated solution.

- Diffusion can take place in gases, liquids and very slowly in solids.

- Osmosis can only take place in the presence of water.

Review questions on Transport across cells

1. Mrs Livingstone is cooking a curry for dinner. The kitchen window is open. Explain how Mr Livingstone knows what is for dinner even before he enters their home.

2. State whether each of the following is an example of osmosis or not.

a) Plant roots absorb water from soil.

b) Carbon dioxide passes from the bloodstream to the air in the lungs.

c) Plants lose water vapour from their leaves into the air.

d) When soft fruit is placed in sugar syrup it gets smaller.

3. Suggest why the smell of petrol diffuses across a garage forecourt more rapidly during the day than at night.

4. Fig 2.RQ.1 shows regions of different concentrations in a plant cell.

a) Name the process by which gases move into and out of cells.

Predict what will happen in each of the following conditions.

b)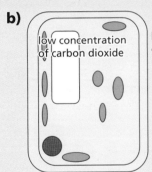

low concentration of carbon dioxide

high concentration of carbon dioxide

c)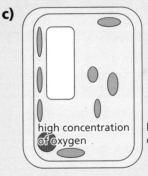

high concentration of oxygen

low concentration of oxygen

FIG 2.RQ.1

5. The concentration of substances in red blood cells is equivalent to 1.7% salt solution. Here is what happens when red blood cells are placed in 4% salt solution.

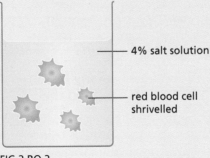

4% salt solution

red blood cell shrivelled

FIG 2.RQ.2

a) Explain what happened.

b) Predict what would happen if the red blood cells were placed in pure water.

Using reverse osmosis to make salt-free water

In Fig 2.SIP.1 equal volumes of water and seawater were placed in the two legs of a U tube. They are separated by a differentially permeable membrane.

As a result of osmosis, water passes into the seawater, causing a difference in the heights of the liquids. This difference is the osmotic pressure.

If an external pressure is applied to the leg of the U tube containing the seawater, as shown in Fig 2.SIP.2, the water molecules move in the opposite direction. This process is called 'reverse osmosis' and is the basis of some desalination processes to obtain fresh water from seawater.

As a science specialist with knowledge of osmosis you have been hired by a local company to investigate how reverse osmosis might be used to make fresh water.

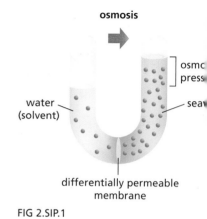

osmosis

osmc
press

water
(solvent)

seav

differentially permeable
membrane

FIG 2.SIP.1

1. You are going to work in groups of 3 or 4 to investigate reverse osmosis. The tasks are:

   • To review those parts of the unit concerned with osmosis.

   • To research reverse osmosis using the Internet and whatever other suitable materials are available.

   • To build a device that can be used to carry out reverse osmosis.

   • To devise a method of assessing how successfully your device works.

   • To investigate the effects of different amounts of external pressure.

   • To compile a report including a PowerPoint presentation in which you explain how your device operates and how well it performed. You should illustrate your report by taking pictures at different stages during manufacture and testing.

   a) Look back through the unit and particularly lessons 2.2–2.4. Make sure that you fully understand the process of osmosis.

   b) Find out what you can about reverse osmosis using whatever sources of information are available to you.

      There are companies in Jamaica who deal with reverse osmosis apparatus. You could try writing to them for information or, if they operate near your school, you could invite them to your school to discuss your project.

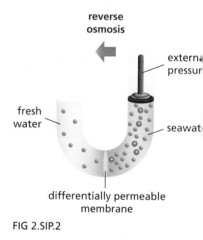

reverse
osmosis

extern;
pressur

fresh
water

seawat

differentially permeable
membrane

FIG 2.SIP.2

c) What sort of device are you going to make? Don't be too ambitious; your target is not to produce a device that will provide the whole school with fresh water but rather a model that could be scaled up after suitable development work.

Your device will need a differentially permeable membrane and some means of applying external pressure. Here is an idea you might try or modify.

d) How are you going to know if your device is working? Clearly if it is you should end up with more fresh water and less sea water. Perhaps you could measure the volume of each before and after the process has taken place?

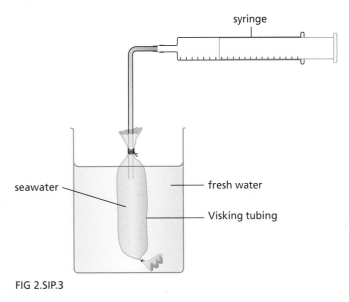

Alternatively, it might be easier to weigh the apparatus containing the fresh water and the remainder of the apparatus containing the seawater before and after? It might help to know that at room temperature the density of fresh water is about 1.00 g/cm³ and the density of seawater is about 1.02 g/cm³.

FIG 2.SIP.3

How are you going to confirm that only fresh water is passing through the membrane?

e) When you are satisfied that your device is actually carrying out reverse osmosis it is time to start experimenting with the effects of pressure. It might appear that the higher the pressure the more fresh water will be produced, but there are experimental constraints. For example, if you are using a Visking bag as your differentially permeable membrane how much pressure can you exert before it bursts?

In order to get quantitative results you will need to devise some method of measuring the pressure exerted. If you are using a syringe to create the pressure you could try calibrating this using a manometer so you can relate the position of the plunger to the pressure exerted.

f) Prepare a PowerPoint presentation in which you describe the structure of your device and explain how you determined whether it was working or not. You should illustrate your account with photographs.

Describe the development work you carried out on the effects of pressure using your apparatus. Show the data you obtained and, if you were able to obtain quantitative data, discuss any mathematical relationship between pressure and rate.

Unit 3: Transport in humans

We are learning how to:
- describe the effect of surface area on the rate of diffusion.

Diffusion and surface area ⟫

The human body relies on diffusion to transport substances into and out of the body. In the lungs oxygen diffuses from the air into the blood while carbon dioxide diffuses in the opposite direction.

In the digestive system, nutrients diffuse out of the small intestine into the blood. Glucose and oxygen diffuse into cells from the blood. Waste products from different metabolic processes diffuse out of the blood to be expelled from the body in urine.

How well the body functions depends on the rate at which diffusion can take place. This, in turn, is linked to surface area.

The **lungs** contain about 300 million tiny air sacs through which gases can diffuse. The total surface area of a pair of adult lungs is about the same as that of a tennis court. Each cluster of air sacs is surrounded by a network of tiny blood capillaries. If all blood capillaries in the lungs were unwound and added together they would stretch for nearly 1000 km. This will give you some idea of the importance of surface area for the diffusion of gases.

The structure of the **small intestine** is similar. The intestine wall is lined with finger-like projections called villi, each of which has its own blood capillaries. These provide a large area for the absorption of nutrients from digested food.

In the activity below you are going to investigate the link between the rate of diffusion and surface area using **agar gel** that has been impregnated with phenolphthalein.

FIG 3.1.1 Air sacs in the lungs

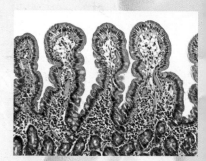

FIG 3.1.2 Cross-section of villi in the wall of the small intestine

Activity 3.1.1

Diffusion in agar gel cubes

Here is what you need:
- Agar gel cubes impregnated with phenolphthalein indicator: 3 cm cube × 2, 2 cm cube × 2, 1 cm cube × 2
- 0.1 m sodium hydroxide solution 200 cm³
- Beaker 250 cm³
- Knife
- Spoon
- Clock or watch.

Fun fact

Agar gel is a nutritious jelly. Microbiologists use agar gel plates to grow colonies of bacteria so they can identify them.

FIG 3.1.3 Colonies of bacteria on an agar gel plate

1. Place about 20 cm³ sodium hydroxide solution in a beaker.

2. Use a spoon to place the cubes in the sodium hydroxide solution.

3. Allow the cubes to remain in the sodium hydroxide solution for 10 minutes and during this time use the spoon to turn them over so all faces are exposed equally at different times.

4. Remove the cubes with the spoon, place them on a paper towel and blot them to dry.

5. Cut each cube in half and make a sketch of each to show how far the sodium hydroxide solution has diffused. You can tell this by the presence of the pink colour.

6. Calculate the volume and the surface area of the three different-sized cubes you used in this activity and write your answers in a table like the one shown in Table 3.1.1.

Length of the sides of the cube/cm	Volume	Surface area

TABLE 3.1.1

7. A 3 cm cube has the same volume as nine 1 cm cubes. How does the surface area of a 3 cm cube compare to the surface area of nine 1 cm cubes?

8. Predict whether diffusion would be quicker throughout a 3 cm cube or nine 1 cm cubes.

Fun fact

Phenolphthalein is an **acid–alkali indicator**. It is colourless in acids, like hydrochloric acid, but turns pink in alkalis like sodium hydroxide solution.

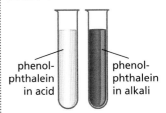

phenol-phthalein in acid

phenol-phthalein in alkali

FIG 3.1.4 Phenolphthalein turns pink in alkalis

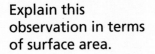

Check your understanding

1. A teaspoon of granulated sugar contains about the same amount of sugar as a sugar cube but it dissolves much more quickly in tea or coffee than a sugar cube.

   Explain this observation in terms of surface area.

FIG 3.1.5

Key terms

lungs part of the body where gases are exchanged

small intestine part of the body where nutrients from digested food are absorbed

agar gel nutrient jelly extracted from seaweed

acid–alkali indicator chemical that is different colours when in acids and alkalis

Structure of the circulatory system

We are learning how to:

- describe the structure of the circulatory system
- explain the function of the heart.

The circulatory system ▶▶

The **circulatory system** is a complex network of blood vessels that delivers blood to every cell in the body. At the centre of the network is the **heart**. The role of the heart is to pump blood around the blood vessels.

The human circulatory system can be described as a **double circulatory system** because it has two circuits or parts to it.

In one circuit, blood passes between the heart and the lungs. While in the lungs, blood absorbs oxygen and releases carbon dioxide and water vapour.

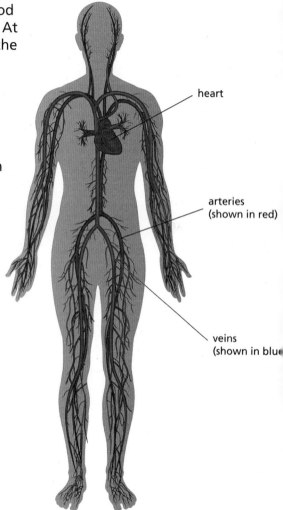

FIG 3.2.1 Human circulatory system

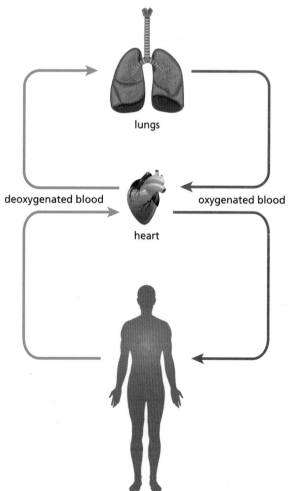

FIG 3.2.2 Double circulatory system

In the other circuit, blood passes between the heart and the rest of the body. Essential substances such as oxygen and glucose are carried to the cells of the body and waste products such as carbon dioxide and urea are removed.

This double system has the advantage that the blood to the rest of the body can be pumped by the heart at a much higher pressure than would otherwise be the case. This ensures that oxygenated blood reaches all parts of the body.

Activity 3.2.1

Finding your way around the circulatory system

Your teacher will help you to organise this activity. It involves making a large map of the human circulatory system.

Here is what you need:

- Pegs
- String
- Large labels for the different parts of the circulatory system fixed on bamboo canes.

Here is what you should do:

1. On the school field mark out the parts of the circulatory system. You should include the heart, the lungs and any other major organs that you know. Place labels inside the organs.

2. Connect the organs using string. You can use different-coloured string to distinguish arteries from veins.

3. Take a walk around the circulatory system and make a note of what is happening at each place you visit.

4. Afterwards, discuss your 'trip' with the other students in your class.

Check your understanding

1. Copy and complete the following sentences.

   a) Blood is pumped around the body by the _____ .

   b) Blood carries the gases _____ and _____
   _____ .

2. What happens in the lungs?

3. What is the advantage of the double circulatory system?

Fun fact

In diagrams of the circulatory system the arteries are traditionally shown in red and the veins in blue. This does not mean that the blood changes colour from red to blue passing through the body.

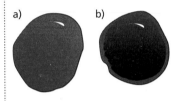

a)   b)

FIG 3.2.3 Samples of: **a)** oxygenated blood **b)** deoxygenated blood

Blood that carries oxygen is described as oxygenated blood and is bright red. Blood that carries blood that has lost its oxygen is called deoxygenated blood and is dark red.

Key terms

circulatory system system that carries blood around the body

heart muscular four-part sac that is at centre of the circulatory system and pumps blood around the body

double circulatory system a circulation system where one circuit is from the heart to the lungs and back, and the other is from the heart to the rest of the body and back

The heart

We are learning how to:

- describe the structure of the heart
- explain the movement of blood through the different chambers of the heart.

The heart >>>

The heart is a muscular organ that pumps blood around the body.

The heart has four chambers. The terms right and left are used as if you are viewing a person's heart from the front of their body. There are two upper chambers – **auricles (atria)** – and two lower chambers – **ventricles**.

The two sides of the heart are separated by a thick muscular wall – the **septum**. This prevents the **oxygenated** blood in the left side from mixing with the **deoxygenated** blood in the right side.

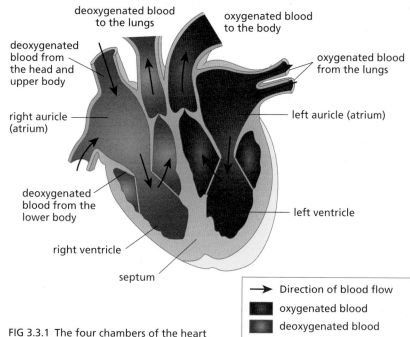

deoxygenated blood to the lungs

oxygenated blood to the body

deoxygenated blood from the head and upper body

oxygenated blood from the lungs

right auricle (atrium)

left auricle (atrium)

deoxygenated blood from the lower body

left ventricle

right ventricle

septum

FIG 3.3.1 The four chambers of the heart

→ Direction of blood flow
oxygenated blood
deoxygenated blood

The flow of blood through the heart is as follows:

body → right auricle → right ventricle → lungs → left auricle → left ventricle → body

Deoxygenated blood returns to the right auricle of the heart from all parts of the body. The blood passes into the right ventricle and is then pumped to the lungs, where it absorbs oxygen and releases carbon dioxide.

The oxygenated blood flows from the lungs into the left auricle. In the heart the blood passes into the left ventricle from where it is pumped to all parts of the body.

Diastole and systole

The heart continues to pump by repeatedly contracting and relaxing every second that you are alive. Heart muscle does not become exhausted and stop as other muscles might.

The moment when the heart muscle relaxes is called the **diastole**. During this time:

- deoxygenated blood from the body enters the right auricle
- oxygenated blood from the lungs enters the left auricle.

The moment when the heart muscle contracts is called the **systole**. During this time:

- the auricles contract, forcing blood into the ventricles
- when full, the ventricles contract, forcing blood out of the heart.

There are valves in the heart to make sure the blood only flows in one direction and cannot flow backwards.

Activity 3.3.1

Examining a pig's or a cow's heart

Here is what you need:

- Pig's/cow's heart
- Scissors
- Scalpel
- Dissecting pins.

Here is what you should do:

1. Carefully examine the outside of the heart. Can you see the coronary blood vessels that supply the heart muscle with blood?

2. Observe where the blood vessels attached to the heart are connected.

3. Cut into the heart lengthwise and identify the four chambers.

4. Look for the valves that prevent blood flowing backwards.

5. Make a sketch or drawing of your dissection.

Check your understanding

1. a) In what order does blood pass through the chambers of the heart?

   b) Which chambers of the heart contain:

      i) oxygenated blood?
      ii) deoxygenated blood?

Something to think about

Some people are born with a congenital defect – a hole in their heart. This means that there is a hole in the septum. This allows deoxygenated blood to pass from the right side of the heart to the left side without passing through the lungs.

As a result, the person's blood does not carry sufficient oxygen around the body so they lack energy and are often tired. This condition can be corrected by surgery to seal the hole.

Key terms

auricles (atria) upper chambers of the heart

ventricles lower chambers of the heart

septum thick muscular wall that separates the two sides of the heart

oxygenated contains oxygen

deoxygenated does not contain oxygen

diastole the moment when the heart muscle relaxes

systole the moment when the heart muscle contracts

Arteries, veins and capillaries

We are learning how to:

- explain the roles of arteries, veins and capillaries
- describe the differences in structure between an artery and a vein.

Arteries and veins

Arteries	Veins
Carry blood away from the heart	Carry blood to the heart
Most carry oxygenated blood (exception is pulmonary artery)	Most carry deoxygenated blood (exception is pulmonary vein)
thick outer wall · narrow diameter · thick layer of muscles and elastic fibres FIG 3.4.1 Cross-section through an artery	fairly thin outer wall · large diameter · thin layer of muscles and elastic fibres FIG 3.4.2 Cross-section through a vein
Have thick walls that include a thick layer of muscle to withstand high pressure of blood leaving heart	Have thinner walls and contain less muscle as blood pressure inside vein is much less than in artery
Inside diameter or **lumen** of an artery is relatively small	Inside diameter or lumen is wider than that of an artery
The blood flow is at high pressure so valves are not needed as there is no chance of back flow.	The blood flowing through veins does not have the benefit of high pressure created by the pumping heart. Instead, it relies on being squeezed through the veins as a result of muscle action. In order to prevent blood flowing backwards, long veins, such as those in the arms and legs, have valves that allow the blood to flow in one direction only. direction of blood flow · open valve · wall of vein · vein squeezed by body muscle · if blood flows back, it fills the pockets, closing the valve FIG 3.4.3 Long veins contain non-return valves

TABLE 3.4.1

Capillaries

Arteries and veins are ideal for transporting blood from one part of the body to another but their walls are far too thick for substances to diffuse into and out of them.

Arteries sub-divide many times, first forming **arterioles** and finally a network of blood **capillaries** so that every cell is supplied with blood.

Blood capillaries are much thinner than human hair and their walls are only one cell thick. This allows substances such as oxygen and glucose to diffuse out into the cells while, at the same time, waste products such as carbon dioxide and urea diffuse out of the cells.

As the blood flows through the capillaries, it provides the substances that the cells require. The capillaries leaving the cells join together to form slightly larger blood vessels called **venules**, which then combine to become veins, carrying the blood back to the heart.

capillary wall is one cell thick

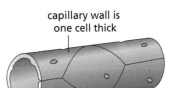

FIG 3.4.4 A blood capillary

Fun fact

The artery that connects the heart with the lungs is the only artery in the body that carries deoxygenated blood. It is called the pulmonary artery.

Activity 3.4.1

Cross-sections of blood vessels

Here is what you need:

- Art paper
- Paints or coloured pens/pencils.

Here is what you should do:

1. Draw a cross-section of an artery, a vein and a capillary next to each other.
2. Use a different colour for each of the materials in the blood vessel walls.
3. Label your blood vessels.

Check your understanding

1. Explain why:
   a) the walls of an artery need to be thicker than the walls of a vein
   b) long veins have valves but long arteries do not
   c) the walls of blood capillaries are much thinner than the walls of arteries or veins.

Key terms

arteries vessels that are part of the circulatory system that carries blood away from the heart

veins vessels that are part of the circulatory system that carries blood to the heart

lumen inside diameter of an artery or vein

arterioles subdivisions of arteries

capillaries vessels that are part of the circulatory system; they are very thin, have walls one cell thick and allow substances to diffuse out into cells

venules larger blood vessels formed by capillaries joining together

Components of the blood

We are learning how to:

- identify the different components of blood
- describe the different roles of the components of blood.

Blood ≫

Blood consists of a liquid part – plasma – and a solid part made up mostly of blood cells. Plasma is about 90 per cent water but also contains some important substances including:

- nutrients obtained by the digestion of food, which are being taken to the cells of the body

- waste products such as urea, which will eventually be excreted from the body

- blood proteins such as antibodies, which help to protect the body from disease, and fibrinogen, which helps the blood to clot

- hormones that coordinate different functions within the body.

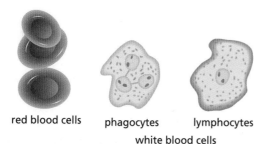

red blood cells   phagocytes   lymphocytes

white blood cells

FIG 3.5.1 Red and white blood cells

The mass of blood cells contains a mixture of both **red blood cells** and **white blood cells**, each with vital roles in our wellbeing. It also contains **platelets**. Platelets look like fragments of red blood cells. They are essential for the clotting process.

Activity 3.5.1

Observing blood cells

Here is what you need:

- Microscope
- Prepared slides showing different blood cells.

Here is what you should do:

1. Place the prepared slides under a microscope and observe first under low power and then under high power.

2. Draw and label examples of the different types of blood cells.

Red blood cells	White blood cells
• Transport oxygen around the body • Contain a pigment called haemoglobin, which contains iron • A shortage of red blood cells causes anaemia – a condition that may be due to insufficient iron in the diet	• Fight any infection that might enter the body • There are two types, **phagocytes** and **lymphocytes**, which fight infection in different ways • Phagocytes enclose bacteria or parts of bacteria into the cell and then digest them and kill them • Lymphocytes release chemicals – antibodies – that destroy bacteria

TABLE 3.5.1

Check your understanding

1. Fig 3.5.3 shows a sample of blood viewed through a microscope.

Which of the following are indicated by A, B, C and D?

a) Platelets
b) Lymphocyte
c) Red blood cell
d) Phagocyte

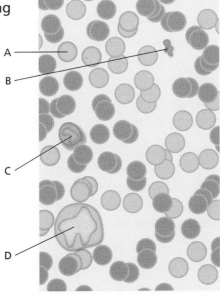

FIG 3.5.3

Key terms

red blood cells cells that transport oxygen around the body

white blood cells cells that fight any infection that might enter the body

platelets small blood particles that are essential for the clotting process

phagocytes cells that enclose bacteria or parts of bacteria into the cell and then digest them and kill them

lymphocytes cells that release chemicals called antibodies that destroy bacteria

Pulse rate

We are learning how to:

- measure a person's pulse rate
- explain the effect of exercise on pulse rate.

The pulse

Pulses of blood are the result of the heart contracting and relaxing. The **pulse rate** is the number of times this occurs in one minute.

Arteries are generally too deep in the flesh to feel the blood pulsing through them but at certain points in the body an artery passes over a bone just under the skin, for example at the wrist, neck, temple and ankle. This makes it possible to feel the pulsing of the blood.

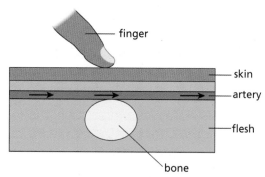

FIG 3.6.1 Feeling a pulse

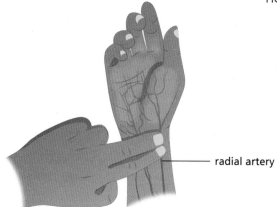

FIG 3.6.2 The radial pulse

The most commonly used point is inside the wrist about two centimetres under the thumb. This is sometimes called the **radial pulse** because it is the result of blood passing along the radial artery to the hand.

An average adult has a pulse rate in the range 60–80 beats per minute. Notice that pulse rate is given as a range. This is because people are all a little bit different. An adult with a pulse rate of 60 beats per minute may be just as healthy as one who has a pulse rate of 80 beats per minute.

Exercise and pulse rate

The circulatory system supplies the body with oxygen and glucose, which are needed to generate energy.

Fun fact

FIG 3.6.3

A giraffe has a large, powerful heart to pump blood up its long neck against the pull of gravity.

When the body becomes more active, perhaps as a result of taking **exercise**, it will need more energy. You might therefore expect the pulse rate to increase as the heart works harder to keep up with demand.

Activity 3.6.1

Investigating the effects of exercise on pulse rate

It is better to work with a partner for this activity.

Here is what you need:

* Stopwatch.

1. Sit at your desk and relax. While you are relaxing, draw a table in your book like the one on the next page.

2. When you and your partner are both relaxed, measure each other's pulse rates at rest by counting the number of pulses per minute.

3. You should both now take five minutes of gentle exercise, such as walking. Your teacher will tell you what to do.

4. After five minutes of gentle exercise, measure each other's pulse rate. Write these values in the table.

5. Your partner and you should now take five minutes of vigorous exercise, such as running. Your teacher will tell you what to do.

6. After five minutes of vigorous exercise, measure each other's pulse rate. Write these values in the table.

FIG 3.6.4 Exercise needs energy

	When I am at rest	When I've taken gentle exercise	When I've taken vigorous exercise
Number of pulses per minute			

TABLE 3.6.1

7. What is the effect of exercise on the pulse rate?

8. Represent the data you have gathered as a bar graph.

Check your understanding

1. Why can a pulse only be felt at certain points of the body?

2. What is the effect of exercise on pulse rate?

Key terms

pulse rate the number of times the heart contracts and relaxes in one minute

radial pulse the pulse inside the wrist just under the thumb

exercise physical activity that increases pulse rate

Review of Transport in humans

- The circulatory system consists of the heart and a network of blood vessels that carry blood to all the cells of the body.

- The heart is a muscular sac that contracts and relaxes throughout a person's life without ever stopping. It consists of four chambers: right auricle, right ventricle, left auricle, left ventricle.

- The diastole is the moment when the heart muscle relaxes. During this time:

  o deoxygenated blood from the body enters the right auricle

  o oxygenated blood from the lungs enters the left auricle.

- The systole is the moment when the heart muscle contracts. During this time:

  o the auricles contract, forcing blood into the ventricles

  o when full, the ventricles contract, forcing blood out of the heart.

- Blood is carried away from the heart in arteries and towards the heart in veins.

- Arteries have thick muscular walls and a small lumen. They must withstand very high pressures as blood is pumped into them from the heart.

- Veins have thinner walls and a larger lumen. The pressure in veins is less than in arteries. Long veins have valves to prevent blood flowing in the wrong direction.

- Arteries divide into arterioles and then into capillaries. The wall of a blood capillary is only one cell thick so substances are able to pass between capillaries and cells. Capillaries combine to form venules. Venules combine to form veins.

- Blood consists of about 90 per cent liquid, which is called plasma, and ten per cent solids, which is mostly blood cells.

- There are different types of blood cells:

  o red blood cells that carry oxygen around the body

  o white blood cells that engulf germs – phagocytes

  o white blood cells that release chemicals that kill germs – lymphocytes.

- A pulse is caused by blood being pumped through arteries by the heart.

- A pulse can be felt at different points of the body where an artery passes over a bone near the surface of the skin. One of the easiest places to feel a pulse is the inside of the wrist.

- Pulse rate is the number of pulses per minute. Pulse rate increases during exercise as the body needs more glucose and oxygen to provide energy.

Review questions on Transport in humans

1. Fig 3.RQ.1 shows a section through a blood vessel.

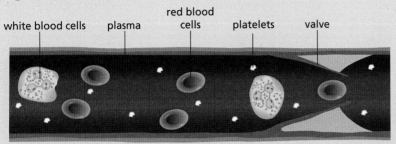

white blood cells   plasma   red blood cells   platelets   valve

FIG 3.RQ.1

   a) Does Fig 3.RQ.1 represent an artery, a vein or a blood capillary? Explain your answer.

   b) In which direction does the blood flow through this vessel? Explain how you know.

2. **a)** Explain why the blood pumped from the heart to the body is brighter red than the blood that returns to the heart.

   b) Fig 3.RQ.2 shows details of the human heart.

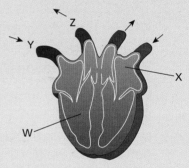

     i) Identify the chambers of the heart marked W and X.

     ii) State where the blood comes from to Y.

     iii) State where the blood goes to from Z.

   c) Some people are born with a hole between the two sides of their heart. Explain why this leaves them feeling weak and lacking in energy.

FIG 3.RQ.2

3. Jessica skipped for 15 minutes. When she stopped, her pulse rate was taken for ten minutes. The results are given in the table.

Pulse rate (beats per minute)	113	100	88	79	73	69	66	64	63	62	62
Time (minutes)	0	1	2	3	4	5	6	7	8	9	10

TABLE 3.RQ.1

   a) Draw a graph of pulse rate on the vertical axis against time on the horizontal axis.

   b) Explain why skipping increases the pulse rate.

   c) What is Jessica's pulse rate at rest?

Making an artificial heart

Some people who have diseased hearts can get a transplant in which they receive a healthy heart from a person who has recently died from other causes, such as a road traffic accident. However, there is a risk that the donor heart may be rejected by the patient's body. Also, there are insufficient donor hearts to satisfy demand. An alternative to a heart transplant could be to replace a diseased heart with an artificial heart.

The Heart Foundation of Jamaica wishes to raise money for research into building an artificial heart. They have asked you to apply your knowledge of the heart to build a simple model which can be used at fund raising meetings. The model needs to demonstrate how a liquid can be pumped from one container into another, in a circuit, using a small electric motor.

1. You are going to work in groups of 3 or 4 to build a simple model that represents the action of the heart. The tasks are:

    *   To review the structure of the heart and the movement of the blood around the circulatory system.
    *   To design a model powered by an electric motor that pumps a liquid from one vessel into another, continuously in a circuit.
    *   To build your model.
    *   To test your model.
    *   To modify your model on the basis of test results.
    *   To demonstrate your model as part of a presentation in which you explain how you went about creating it.

    a) Look back through the unit and make sure you understand how the heart functions in terms of pumping blood around the body.

    b) In essence you are being asked to create a model of single circulation. This can be between the right ventricle and left auricle (via the lungs), or the left ventricle and the right auricle (via the body).

    What are you going to use for your containers? They can be open, such as two beakers, or you might decide to use empty plastic bottles.

    What are you going to use to carry the liquid? It might be better to use clear plastic tubing rather than rubber tubing so the liquid can be observed.

    You can make some lifelike 'blood' with water and a few drops of red food colouring.

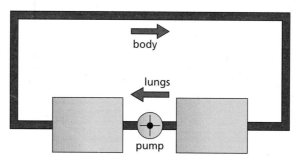

FIG 3.SIP.1 Single circulation

c) The pump is going to provide the kinetic energy needed for your model. How are you going to make a pump out of an electric motor?

One solution might be to attach a circular rotor to the end of the motor. This would have blades twisted at an angle like a propeller. You could build a waterproof housing around it from an old plastic bottle.

You could build a more lifelike version in which the motor raised and lowered a piston. This would give you pulses in the same way the heart beats, however you would have to devise some valves to ensure the liquid flowed in one direction only.

Take pictures as you build your model. You can use these to illustrate your presentation and make it more interesting.

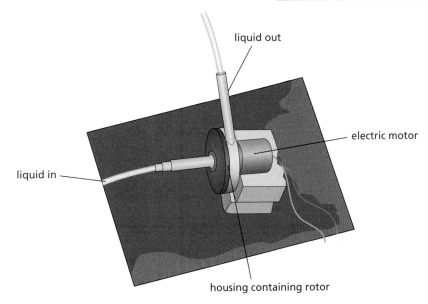

FIG 3.SIP.2 A simple pump

d) Once you have built your model you need to test it. Run it for a few minutes and make sure the liquid is actually circulating.

To improve performance you might:

- Use silicone sealant to ensure the model is watertight.
- Add a potentiometer (variable resistor) to control the speed of the motor. Your teacher will help you with this.
- Reduce the diameter of the tubing so the liquid circulates more quickly.

e) Your final task is to give an illustrated presentation of how you built your model. This should include a demonstration of how it works.

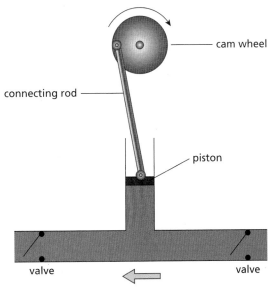

FIG 3.SIP.3 A more realistic pump you could build

Unit 4: Transport in plants

We are learning how to:

• identify the different parts of a flowering plant
• describe the functions of the roots, stem, leaves and flowers of a plant.

Fig 4.1.1 shows the four main parts of a flowering plant. Each of these parts continually grows as the plant develops.

The four parts of a plant labelled in the diagram have very different functions.

Roots >>

The **roots** are the part of a plant that we don't normally see because they are buried in the soil. Unlike all the other parts of a plant, roots are not green because they do not contain the pigment chlorophyll.

Roots have the following important functions:

- They anchor the plant in the soil, which prevents it from being blown about in the wind.
- They absorb water containing dissolved minerals from the soil. This is sometimes called soil water. Both water and the minerals it contains are essential for the plant to flourish.
- They are a place where the plant can store nutrients. This is stored as starch.

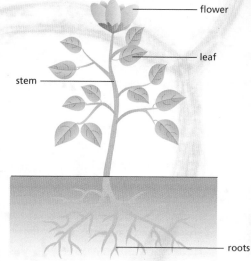

FIG 4.1.1 Main parts of a flowering plant

Stem >>

The **stem** is the part of the plant that joins the roots to the leaves and the flowers. It has the following important functions:

- It supports the parts of the plant that are above the ground. These are the leaves and the flowers.
- It allows the movement of water and minerals up from the roots to the other parts of the plant, and the movement of nutrients from the leaves to the roots for storage. You will learn more about this in a later lesson.
- In climbing plants, like beans, the stem provides attachment by growing in a spiral around another plant or a stick.
- Green stems are able to carry out photosynthesis to make sugars but this process is more important in the leaves.

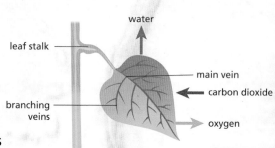

FIG 4.1.2 Veining in a leaf

The **leaves** are the most obvious part of a plant. It is the part we first see when we look at a plant. Each leaf has a network of veins that allow water and nutrients to be transported into and out of the leaf cells.

Leaves are green because they contain a green pigment called chlorophyll which is able to trap energy from sunlight. This energy is used by the plant to convert carbon dioxide and water to the sugar glucose, and oxygen is also formed.

The leaves have the following important functions:

- They make nutrients by the process of photosynthesis.
- They are involved in gaseous exchange.
- They also lose water vapour in a process called transpiration.
- They store some nutrients in the form of starch.

Fun fact

Some plants, like begonias, have leaves that are not green or not totally green. You might think they don't contain chlorophyll. This isn't the case. These leaves do contain chlorophyll but they also contain other pigments that mask the green colour.

Flowers ⟫⟫

Flowers are the organs for sexual reproduction of a flowering plant.

Activity 4.1.1

Observing the different parts of a plant

You should work in a group for this activity.

1. Examine carefully each of the four parts of a sample plant.
2. Make a labelled drawing of the four parts you have examined.
3. On your drawings, annotate the labels to indicate the functions of each part.

Check your understanding

1. Fig 4.1.3 shows a green plant.

   a) Water passes through the plant from the container. In the correct order, list the parts of the plant through which the water passes.

   b) What change took place over the five days that the plant was left in the container?

   c) Suggest a reason for this change.

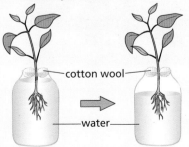

FIG 4.1.3

cotton wool

water

Key terms

root part of plant in the ground that absorbs water and nutrients

stem part of plant that joins roots to leaves

leaf part of plant where food is made, gases are exchanged and water is lost

flower plant organ of sexual reproduction

Plant roots

We are learning how to:

- describe the structure of plant roots
- explain how plant roots function.

Roots ▶▶

The roots are the part of the plant that is in the soil. They absorb water containing dissolved minerals from the soil.

FIG 4.2.1 **a)** Fibrous root   **b)** Tap root

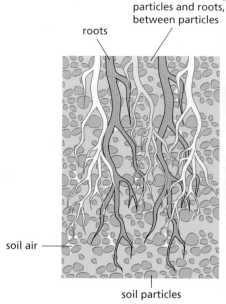

Tap root   Fibrous root

FIG 4.2.2

A plant might have a **fibrous root** which divides into thinner and thinner roots, such as a grass, or a **tap root**, which is a large swollen root with smaller roots growing from it, like a carrot.

Plant roots are ideally suited to absorbing water from the soil. They divide many times so there is a large surface area available for absorption.

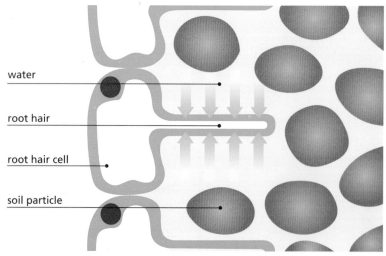

water

root hair

root hair cell

soil particle

FIG 4.2.3 Plant root cells

water film covering particles and roots, between particles

roots

soil air

soil particles

FIG 4.2.4 Plant roots divide many times

Plant roots consist of plant root cells. Each cell has a **root hair** that grows out into the gaps between soil particles, increasing the surface area of the roots even more. Water containing minerals is absorbed through the root hairs.

Activity 4.2.1

Examining roots

Here is what you need:

- Potted plant or plant collected from the school garden

- Hand lens.

Here is what you should do:

1. Carefully remove a plant from the soil so that some soil remains around the roots.

2. Look carefully at the roots. Notice how they are spread out to provide the plant with a firm anchor but also to absorb water and nutrients from different parts of the soil. Notice how the roots divide a number of times to finally produce very thin roots.

3. Knock away some of the soil and examine the area immediately around the roots. Can you see any root hairs?

4. Make some drawings of the roots showing different features.

Check your understanding

1. Fig 4.2.5 shows two different types of root system.

   a) Which root stores more nutrients?

   b) In what ways are the root systems designed to gather water and nutrients efficiently?

tap root system   fibrous root system

FIG 4.2.5

Key terms

fibrous root type of root that divides many times, producing a network of tiny roots

tap root main root from which small roots arise

root hair finger-like projection from the outermost cells of the root which increases their surface area for absorption

Xylem and phloem

We are learning how to:

- identify the xylem and phloem in a plant stem.

Xylem and phloem

You have already seen how water containing dissolved minerals is absorbed by the roots and passes to the other parts of the plant. However, there is also a flow of substances in the plant.

The leaves use energy from sunlight to make food in the form of glucose. The glucose is then distributed to the other parts of the plant either to provide cells with energy, or to be stored as starch.

The plant therefore needs two vessels to transport dissolved substances:

- The **xylem** transports water and water-soluble minerals from the soil to the different parts of the plant.

- The **phloem** transports glucose, proteins and other organic chemicals within the plant.

In the root of a plant the xylem and phloem together form a vascular cylinder.

Within the vascular cylinder the xylem forms the centre, which is surrounded by the phloem.

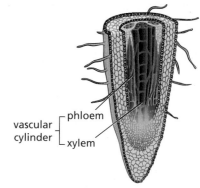

FIG 4.3.1 Vascular cylinder in the plant root

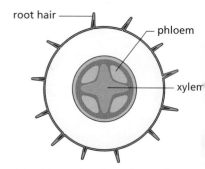

FIG 4.3.2 Section through vascular cylinder in the plant root

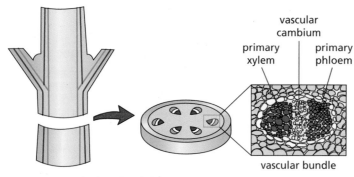

FIG 4.3.3 Vascular bundle in plant stem

A cross-section of a plant stem shows that the vascular cylinder forms a series of **vascular bundles**. These are arranged symmetrically around the stem but the pattern is different in different plants. Each vascular bundle consists of xylem cells towards the centre of the stem and phloem cells towards the outside. These cells are separated by **vascular cambium** cells.

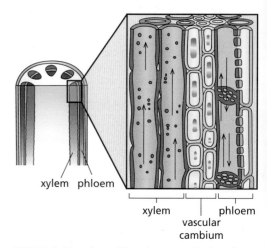

FIG 4.3.4 Direction of flow in a vascular bundle

The flow through the xylem vessels is upwards from the roots towards the other parts of the plant. The flow through the phloem may be in either direction as substances are carried between the different parts of the plant.

Activity 4.3.1

Arrangements of vascular bundles

Here is what you need:

- Poster-sized sheet of paper
- Different coloured card × 2
- Scissors
- Glue.

Here is what you should do:

1. Cut 20 discs of two different colours and sizes from coloured card. Ten will represent the xylem and ten the phloem.

2. Arrange the discs to show how the xylem and phloem are arranged in the stem of a plant.

3. Use the materials provided to find a creative way to show the arrangement of xylem and phloem in the root.

4. When you are satisfied with your arrangement, glue the discs in place so your poster can form part of a classroom display.

Check your understanding

1. Fig 4.3.5 shows a section across a plant stem.

   a) Name the parts labelled A, B and C.

   b) What is carried in:

      i) A?   ii) B?

   c) In which direction is the flow through:

      i) A?   ii) B?

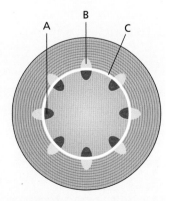

FIG 4.3.5

Key terms

xylem vessels carrying solution of mineral salts from the soil to different parts of the plant

phloem vessels carrying solution of nutrients and other chemicals around the plant

vascular bundle structure containing xylem and phloem vessels

vascular cambium cells between the xylem and phloem in a vascular bundle

Movement of substances along the phloem and the xylem

We are learning how to:

• describe the roles of the xylem and phloem in a plant stem.

Movement of substances ⟩⟩

Both the xylem and the phloem appear to be long thin tubes, like narrow drinking straws, but their structures are very different.

Structure of the phloem ⟩⟩

Phloem tissue consists of two main types of cells, **sieve elements** and companion cells, together with other types of cells that make up the structure.

Sieve elements are long narrow cells that join together to form a sieve tube. At the ends of each sieve element there is a sieve plate. Sieve plates are porous and therefore allow substances to flow between cells along the sieve tube.

Sieve elements have no nuclei. They do, however, have thick, rigid cell walls made of cellulose. These are needed to withstand the hydrostatic pressures that bring about the flow through the phloem.

Sieve elements would not be able to function without companion cells. The companion cells make the movement of materials through the sieve plates possible.

Substances are able to move in both directions through the sieve tubes. Movement is the result of hydrostatic pressure from the xylem.

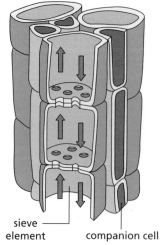

sieve element   companion cell

FIG 4.4.1 Sieve elements and companion cells in the phloem

Structure of the xylem ⟩⟩

The xylem consists of stacks of dead cells joined together to form tubules.

Since the xylem doesn't contain living cells it is not possible for the solution of mineral salts to pass up the plant by osmosis. Instead, the solution of mineral rises up through the xylem by **capillarity**. This occurs when liquids are in tubes with very small diameters.

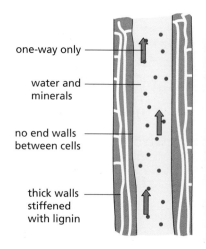

one-way only

water and minerals

no end walls between cells

thick walls stiffened with lignin

xylem vessel

FIG 4.4.2 Stack of dead cells making up a xylem vessel

Activity 4.4.1

Examining the tubes that carry water up a stem

Here is what you need:

- Solution of red ink

- Stick of celery

- Sharp knife.

Here is what you should do:

1. Cut off the bottom part of a fresh celery stalk and stand it in red ink solution at the start of a lesson.

2. Near the end of the lesson take the stalk out of the ink solution and cut it across the middle with a sharp knife.

3. Examine the cut part of the celery. Look for the red dots where the ink has been drawn up through the stem.

4. Using your knowledge of stems what do you think the red spots represent?

5. Make a drawing to show the arrangement of the red spots.

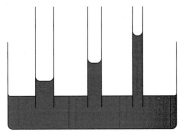

FIG 4.4.3 The narrower the tube, the greater the effect of capillarity

Check your understanding

1. Decide whether each of the following describes the xylem, the phloem or both the xylem and the phloem.

   a) Carries substances dissolved in solution.

   b) Substances only travel in two directions.

   c) Substances can only travel along by capillarity.

   d) Contains sieve elements.

   e) Consists of stacks of dead cells.

2. a) What is capillarity?

   b) Describe an experiment you could carry out to demonstrate capillarity using ink solution and a glass tube with a very narrow bore.

   c) Draw a diagram to show the result of your experiment.

   d) Why is capillarity important for the movement of substances in a plant?

> **Fun fact**
>
> Cut flower stems can absorb solutions that will alter the natural colour of the flowers.
>
>
>
> FIG 4.4.4 Have you ever seen blue roses in a garden?
>
> Florists use this trick to make unusual floral displays.

Key terms

sieve tube a tube composed of sieve elements in the phloem

capillarity movement of a liquid up narrow tubes

Transpiration

We are learning how to:

- describe the process of transpiration
- explain why transpiration is important to a plant.

Transpiration ⟫

Absorbing minerals from the soil is essential to the growth of a plant but there is a limit to how much water plants can store. What happens to all of the water absorbed by a plant? The absorption of water and minerals from the soil is only one part of a bigger process called **transpiration**.

Transpiration is the movement of water through a plant. Water is absorbed by the roots and passes through the plant. It is eventually lost by evaporation, mainly through the leaves. This is sometimes called a transpiration stream.

If the stem and leaves of a plant are sealed in a polythene bag and left for 24 hours, condensation forms inside the bag. The water is not coming from the soil but from the leaves of the plant.

As a result of transpiration, water is continually lost from the surface of leaves and so more water is drawn up through the xylem to replace it. In turn, soil water containing minerals is absorbed by the plant roots to replenish the xylem. This is called transpiration pull.

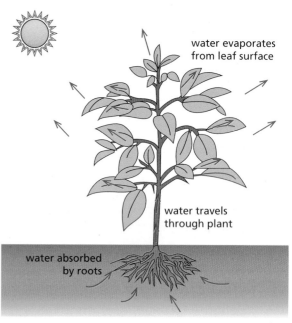

water evaporates from leaf surface

water travels through plant

water absorbed by roots

FIG 4.5.1 Transpiration through a plant

Activity 4.5.1

Investigating water loss from both surfaces of a leaf

Here is what you need:

- Plant with green leaves
- Blue cobalt chloride paper
- Adhesive tape
- Scissors.

Here is what you should do:

1. Cut out two pieces of blue cobalt chloride paper about the same size as a leaf.

FIG 4.5.2 Transpiration causes condensation inside a polythene bag

2. Using adhesive tape, attach the cobalt chloride paper to either side of a leaf. The tape should extend beyond the edge of the leaf so that the front and back tapes can be stuck together (Fig 4.5.3).

3. Make sure the plant is well watered and leave it for several hours.

4. Observe the colour of the cobalt chloride paper covering the upper surface and the lower surface of the leaf.

5. Blue cobalt chloride paper turns pink in the presence of water. Which paper has turned more pink? What does this show?

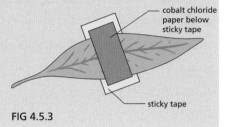

cobalt chloride paper below sticky tape

sticky tape

FIG 4.5.3

In addition to allowing the uptake of minerals, transpiration has other important benefits for a plant:

- Transpiration allows the plant to get rid of excess water.
- Transpiration creates hydrostatic pressure, which allows the plant substances to be transported around the plant.
- When water **evaporates** from the surface of a leaf it removes heat energy. A plant can increase the rate at which water is lost from the leaves in order to cool its leaves and prevent them being damaged by heat from the Sun.

Fun fact

Although carnivorous plants like the Venus flytrap absorb water through their roots just like other plants, they don't rely on soil water for minerals.

FIG 4.5.4 A Venus flytrap

Carnivorous plants obtain nutrients by trapping and digesting insects.

Check your understanding

1. In an experiment carried out to demonstrate transpiration a polythene bag was placed over the leaves and stem of a plant. The open end of the bag was sealed around the stem.

   a) Predict how the plant and bag will appear after 24 hours.

   b) Explain your prediction.

   c) Why was the open end of the bag sealed around the plant stem and not the top of the pot?

polythene bag

FIG 4.5.5

Key terms

transpiration movement of water through a plant

evaporate when a liquid changes to vapour below its boiling point

Review of Transport in plants

- A plant consist of four parts: roots, stem, leaves and flowers.

- The roots anchor the plant in soil and absorb water containing dissolved minerals.

- The stem connects the roots with the leaves.

- The leaves use energy from sunlight to make food.

- The flower is the organ of sexual reproduction.

- A plant may have a fibrous root or a tap root.

- Tap roots contains stores of food in the form of starch.

- Fibrous roots divide many times to give a network of roots which has a large surface area.

- Root cells have finger-like projections called root hairs that increase their surface area for absorption.

- Substances move around a plant via the xylem and phloem.

- The xylem transports water and water-soluble minerals from the soil to the different parts of the plant.

- The phloem transports glucose, proteins and other organic chemicals within the plant.

- In the root the xylem and phloem form a vascular cylinder.

- In the stem the xylem and phloem form vascular bundles.

- Substances pass only one way up the xylem.

- Substances may pass either way in the phloem.

- Substances rise up the xylem by capillarity.

- Transpiration is the movement of water through a plant.

- Plants absorb water through the roots and mostly lose it through the leaves.

- Transpiration cools the leaves so they are not damaged by heat from the Sun.

Review questions on Transport in plants

1. Fig 4.RQ.1 shows the outline of a plant.

   a) Identify parts A to D.

   b) In which part:

      i) is most water lost?

      ii) is water absorbed?

   c) What is the function of part A?

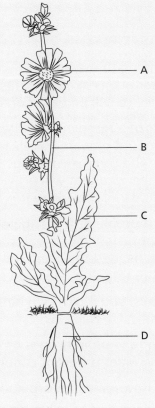

FIG 4.RQ.1

2. a) In a plant what is the function of:

      i) the xylem?

      ii) the phloem?

   b) Draw a labelled diagram to show how the xylem and phloem form vascular bundles within a plant stem.

3. The plants in Fig 4.RQ.2 were left in a warm place for 24 hours.

   a) Predict what will happen.

   b) Explain your prediction.

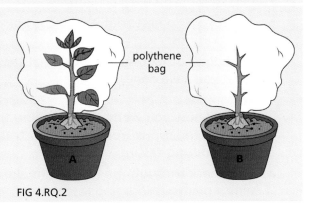

polythene bag

FIG 4.RQ.2

4. a) What colour does blue cobalt chloride paper becomes when it is damp?

   b) How can cobalt chloride paper be used to demonstrate that a plant loses more water through the lower surface of a leaf than the upper surface?

Providing pot plants with adequate water

A company is planning to market a device that will automatically water plants in a pot while the owners are away from home. The device will be based on a 20 cm diameter pot.

The device consists of a reservoir of water and a mechanism by which a certain amount of water is released into the growing medium each day. The company intends to offer the device in two sizes: a smaller size that will provide water for a week, and a larger size that will provide water for two weeks.

FIG 4.SIP.1 Self-irrigation

However, there is a problem. The engineer designing the device does not know how much water a plant loses each day. This information is needed to determine the sizes of the reservoirs. The more water a plant loses, the bigger the reservoir will have to be.

As a science specialist you have been hired by the company to investigate and make recommendations.

1. You are going to work in groups of 3 or 4 to investigate how much water a typical plant loses each day under different conditions. The tasks are:

   - To review how plants absorb and lose water.
   - To devise a method of measuring the amount of water lost by a potted plant in a day.
   - To investigate whether external conditions have any effect on the rate at which a potted plant loses water.
   - To determine an average value for the amount of water lost by a potted plant in one day.
   - To test the validity of your value by watering a potted plant with this amount of water each day over a week.
   - To make recommendations to the company for reservoir sizes based on your observations.
   - To suggest other research work that the company might undertake that builds upon what you have done.
   - To compile a report including a PowerPoint presentation in which you should explain how you went about solving the problem and how you arrived at your recommendations. You should illustrate your report by taking pictures at different stages during testing.

   a) Look back through the unit and make sure that you understand the mechanisms by which plants absorb and lose water.

   b) How are you going to determine how much water a potted plant loses in a day? It might be useful to remember that 1 cm³ of water has a mass of 1 g so that water loss can be related to loss of mass.

You may decide to water the plant first thing in the day, let it drain and then weigh it. You could then weigh it first thing the following day to find out how much water has been lost.

c) The weather changes from day to day. What external conditions might affect the loss of water from a potted plant? Perhaps plants lose more water on warmer days than on cooler days. How are you going to find out? You might experiment by placing a potted plant in a sunny position one day and a shady position the next. Are there other external conditions that need to be considered? How are you going to record the data you obtain from your experiments?

FIG 4.SIP.2 Weighing a potted plant

d) Over a period of a week or a fortnight the weather and the amount of water lost by a potted plant each day will vary. Basing the size of the reservoir on the amount of water lost on the hottest day might lead to an overestimate of what is needed, but basing the amount on the coolest day might lead to an underestimate and the plant might dry up and die.

You need to find an 'average' amount of water loss over a week and a fortnight. How will you do this?

e) People buying this device won't be very pleased if it doesn't provide enough water and their plants die. How are you going to test whether your average amount is sufficient? What will you do if it turns out not to be so?

f) How are you going to use your results to recommend the size of reservoir needed for a device that waters a potted plant for a week, and one that waters a potted plant for a fortnight?

g) Do you think that the development work you have carried out provides the company with sufficient information to go ahead and manufacture their product or do you think there are other factors that might be important? For example, do all types of plants lose the same amount of water each day? Do plants in clay pots lose water at the same rate as plants in plastic pots?

h) Prepare a PowerPoint presentation in which you describe what you did in order to find the average amount of water lost by a potted plant each day. You should illustrate your account with photographs.

Show the data you obtained and explain how you interpreted this data in order to make recommendations about reservoir size.

Discuss other issues that you think the company should consider and perhaps research before finally deciding on the sizes of the reservoirs needed for their device.

Unit 5: Static electricity

We are learning how to:

- describe the structure of an atom
- describe some properties of sub-atomic particles.

Structure of the atom ≫

The atom

Atom comes from the Greek word 'atomos', which means indivisible. The idea of a particle which could not be divided into anything smaller was first proposed over 2000 years ago by an Ancient Greek called Democritus. Atoms are the building blocks from which all things are made.

An **element** is a substance that cannot be made into any simpler substance. All of the atoms of one element are similar in structure, but different from the atoms of any other element.

There are 94 naturally occurring elements and many will be familiar to you. These elements can be divided into three groups according to their properties:

- Metals like aluminium and copper
- Non-metals like oxygen and carbon
- Metalloids like silicon.

a) Aluminium

b) Carbon

c) Silicon

FIG 5.1.1 Some common elements

Sub-atomic particles ≫

You should recall that an atom consists of a nucleus containing **protons** and **neutrons**. Surrounding the nucleus are layers of **electrons**. Table 5.1.1 summarises the properties of these three **sub-atomic particles**.

Sub-atomic particle	Relative mass	Relative charge	Position in the atom
Proton	1	+1	In the nucleus
Neutron	1	0	In the nucleus
Electron	0	−1	Around the nucleus in shells

TABLE 5.1.1

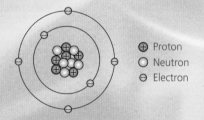

⊕ Proton
○ Neutron
⊖ Electron

FIG 5.1.2 Structure of an atom

Activity 5.1.1

Building a model of an atom of carbon

You should work in a small group for this activity.

Here is what you need:

- Small polystyrene balls or similar spheres, 12 large and 6 small

- Thin wire
- Paints
- Glue
- Thread.

1. Decide on colours to represent protons, neutrons and electrons.

2. Paint six of the balls with each colour. Remember that the electrons are smaller.

3. Glue the protons and neutrons together to form the nucleus of your atom.

4. Make a loop of wire a little larger than your nucleus and thread two of the electrons onto the wire. Keep them in position on the loop with glue.

5. Make a second loop a little larger than the first, then thread the remaining four electrons onto it. Hold them in position with glue.

6. Attach a piece of thread with glue, firstly to the nucleus, and then loop it around the rings of electrons and hold it in position with glue.

7. Find a suitable place in your classroom to suspend your atom.

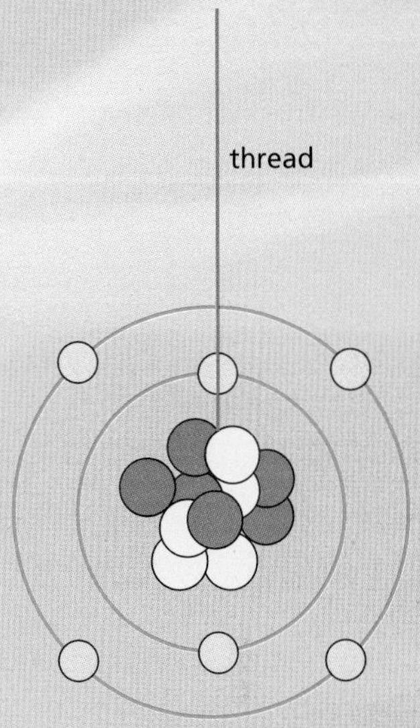

FIG 5.1.3 Model of a carbon atom

Check your understanding

1. Copy and complete the following sentences.

   a) Atom comes from a Greek word that means ....

   b) An .... is a substance that cannot be changed into simpler substances.

   c) An .... is the smallest particle of an element that can exist.

   d) .... particles are particles found inside an atom.

2. Answer the following questions about sub-atomic particles.

   a) Which particle has a relative mass of approximately 0?

   b) Which two particles have the same mass?

   c) Which two particles carry equal but opposite charges?

   d) Which particles are found in the nucleus of an atom?

Key terms

atom the smallest particle of a substance that can take part in a chemical reaction

element substance that cannot be made into simpler substances

proton positively charged particle in the nucleus of an atom

neutron neutral particle in the nucleus of an atom

electron negatively charged particle surrounding the nucleus of an atom

sub-atomic particles are particles found within the atom

Investigating static electricity

We are learning how to:

- explain what static electricity is
- describe some of the properties of static electricity.

Static electricity >>>

The word **static** means not moving. We normally think about electricity as passing very quickly along metal wires but, under the right conditions, electrical charge can build up on objects made of materials like glass and plastic.

All matter is composed of tiny particles called atoms. **Electrical conductors** like metals are able to conduct or transfer charge because the outermost electrons of each atom are not in fixed positions but are delocalised and able to move within the conductor. It is these mobile electrons which act as charge carriers.

Electrical insulators like glass and plastics do not have mobile electrons. Electric charge builds up on the surface of the insulator because it cannot flow away. The insulator can only be discharged by connecting it to Earth with a conductor.

FIG 5.2.1 Static electricity can build up on our bodies

Charging by friction >>>

You may have noticed that when you use a plastic comb on your hair it sometimes crackles. Static electricity is created by **friction** when you rub your hair with the comb.

Under certain conditions charge even builds up in the atmosphere. When cold and warm bodies of air rub over each other friction causes them to become charged. Eventually the charge becomes so great it passes to Earth in a great spark. We call this **lightning**. This discharge of static electricity is often accompanied by an explosion we call thunder.

FIG 5.2.2 Lightning is the discharge of static electricity to Earth

Charging with static electricity >>>

Any insulator can be charged with static electricity by rubbing it with another insulator. In the laboratory, rods of glass or plastic are charged by rubbing them on fabrics such as wool or silk.

When the two insulators are rubbed together, electrons are transferred from one substance to the other. As electrons carry a negative charge, the insulator that loses electrons

FIG 5.2.3 A charged plastic ruler attracts small bits of paper

becomes positively charged while the insulator that gains electrons becomes negatively charged.

Materials charged	Material positively charged	Material negatively charged
Glass rod rubbed with silk	glass	silk
Ebonite rod rubbed with fur	fur	ebonite
Perspex ruler rubbed with a duster	Perspex	duster
Plastic comb rubbed with wool	wool	plastic comb

TABLE 5.2.1 Charging by friction

Fun fact

The ancient Greeks noticed that rubbing amber on wool gives it unusual properties. It is able to pick up small objects likes hairs, threads, pieces of feathers and even dry leaves.

Activity 5.2.1

Investigating some effects of static electricity

Here is what you need:

- Plastic comb (must be clean and not greasy)
- Glass rod
- Piece of wool
- Rubber balloon
- Small pieces of tissue paper.
- Piece of silk

1. Pull a comb through your hair ten times. Place the comb near some small pieces of paper. Describe what you observe.

2. Rub a glass rod with a piece of silk.

3. Turn on a tap so there is a very thin column of water flowing. Bring the charged glass rod close to the column of water but don't touch it. Describe what you observe.

4. Blow up a balloon. Rub the balloon on a piece of wool or a woollen jumper.

5. Place the rubbed part of the balloon against the wall and then carefully move your hands away. Describe what you observe.

Check your understanding

1. Fig 5.2.4 shows what happened when a positively charged Perspex rod was placed near to a stream of water.

Explain this behaviour.

FIG 5.2.4

Key terms

static not moving

friction force that opposes motion

electrical conductor material that conducts an electric current

electrical insulator material that does not conduct an electric current

lightning movement of electrical charge between the sky and the ground

71

Uses of static electricity

We are learning how to:

- describe some uses of static electricity.

Electrostatic charging ⟫

Electrostatic charging has a number of important applications.

Spray painting

In manufacturing processes where objects are automatically spray painted, the paint and the object are oppositely charged.

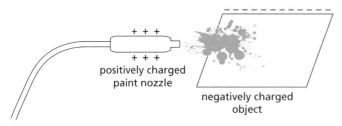

positively charged paint nozzle

negatively charged object

FIG 5.3.1 Spray painting

As the paint passes through the nozzle it becomes charged. The paint droplets carry the same charge and repel each other. This ensures a broad spray of paint.

The paint droplets carry the opposite charge to the object so they are attracted to it. The object is sprayed efficiently using the minimum amount of paint, with very little lost to the surroundings.

Photocopiers

Electrostatic charge provides a means of copying documents.

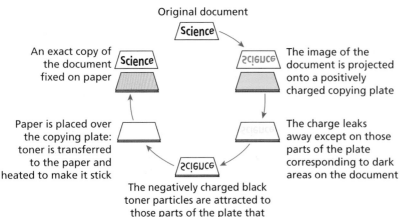

Original document

An exact copy of the document fixed on paper

The image of the document is projected onto a positively charged copying plate

The charge leaks away except on those parts of the plate corresponding to dark areas on the document

Paper is placed over the copying plate: toner is transferred to the paper and heated to make it stick

The negatively charged black toner particles are attracted to those parts of the plate that remain positively charged

FIG 5.3.2 Photocopying

> **Fun fact**
>
> Some people believe that a build-up of electrostatic charge on a car in which they are travelling causes travel sickness.
>
> They attach an **'antistatic'** strip which connects the car body to the ground.

Key terms

electrostatic static electrical charge

antistatic prevent build-up of static

When a document is scanned on a photocopier the image is projected onto a positively charged plate. Where light falls on the plate the charge leaks away. This leaves a charged area on the plate corresponding to the dark part of the document.

Negatively charged black (or colour) toner powder is then attracted to those parts of the plate that are still charged. This produces a mirror image of the document on the plate.

Paper is placed over the plate. The toner is transferred to the paper and fixed by heating. The result is a perfect copy of the original document.

Activity 5.3.1

Making an electrostatic separator

Here is what you need:

- Balloon
- Duster or woolly jumper
- Mixture of salt and finely ground pepper
- Open dish or plate.

1. Sprinkle the mixture of salt and pepper on the plate.

FIG 5.3.3

2. Blow up the balloon and tie the end.

3. Rub the balloon with a duster or on a woolly jumper to charge it.

4. Place the charged balloon just above but not touching the plate.

5. Describe your observations.

6. Explain how the charged balloon is able to separate the salt and pepper.

7. Can you suggest an industrial process in which static electricity may bring about a separation of particles?

Check your understanding

1. Insecticides can be spread over large areas when spread from an aeroplane. The insecticide leaves the aeroplane through charged nozzles.

   a) Describe what happens to the insecticide as it passes through the nozzle.

   b) Explain why charging the nozzle allows the insecticide to be used more effectively and efficiently.

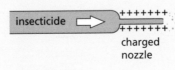

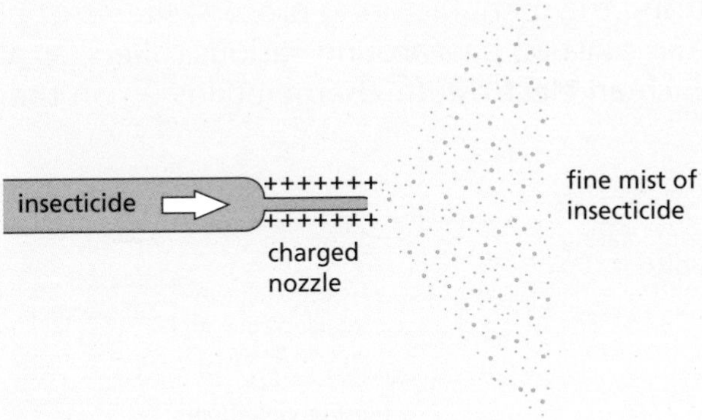

FIG 5.3.4

Problems caused by static electricity

We are learning how to:

- describe some problems caused by static electricity.

Problems caused by static electricity >>>

The build-up of electrostatic charge on insulating materials or materials not connected to Earth creates some serious hazards.

Lightning

Under certain weather conditions, massive amounts of electrostatic charge may build up in the atmosphere and finally discharge to Earth as lightning. Lightning tends to discharge through objects sticking up from the ground like trees, tall chimneys and towers.

Lightning conductors are attached to many buildings to prevent or minimise damage in the event of a lightning strike. The conductor consists of thick copper or brass wire that provides the electric charge with an easy passage to Earth.

FIG 5.4.1 Lightning conductors prevent damage to buildings

Powders in pipes

Many industrial processes involve the movement of powders such as coal dust, grain, chocolate powder and flour, along pipes. In years gone by, it was not appreciated that friction between a powder and a pipe created electrostatic charge.

In a building such as a flour mill, where the air contained combustible particles, the small spark formed when static charge was earthed was sometimes enough to trigger an explosion. In modern factories, pipes are earthed to prevent any build-up of static charge.

FIG 5.4.2 Remains of a flour mill after an explosion and fire

Polythene manufacture

During the manufacturing process, sheets of polythene (and other plastics) pass around various rollers, as a result of which an electrostatic charge builds up on the polythene.

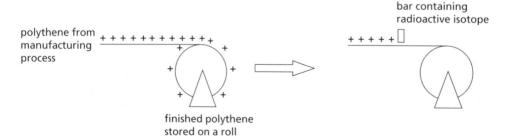

FIG 5.4.3 Static electricity builds up on polythene during manufacture

If the finished polythene remains charged, it is difficult to store on a roll because each new layer is repelled by the previous layer on the roll. The result is that the polythene sheet tends to slew sideways on the roll.

The electrostatic charge is removed by placing a bar containing a **radioactive isotope** across the polythene, just above but not touching it. The isotope causes the air above the polythene to become charged, and this neutralises the charge on the polythene.

FIG 5.4.4

Activity 5.4.1

Assessing risk when transferring fuels

You should work in a group on this activity.

The build-up of electrostatic charge occurs with liquids in the same way as it does when particles move along pipes. This is a particular danger when the liquid is highly flammable, like petrol, diesel or aviation fuel.

If static charge built up while a car was being filled with fuel, a spark between the metal nozzle and the car would be enough to set the fuel alight.

Your task is to research the problems created by static electricity when transferring fuel. You should:

- Identify how electrostatic charge might build up.

- Explain what hazards result from the build-up of electrostatic charge.

- Describe what safety procedures are adopted to counter the hazards.

Key terms

lightning conductor
metal strip to protect buildings from lightning damage

radioactive isotope
substance that emits radiation

Check your understanding

1. During the manufacture process, artificial fertiliser powder acquires an electrostatic charge. Fig 5.4.5 shows what happens when bags full of fertiliser powder are sealed.

    a) How does the polythene bag become charged?

    b) Why does the electrostatic charge make it difficult to close the bags?

    c) A radioactive source charges the air around it. Explain how this makes it easier to close the bags.

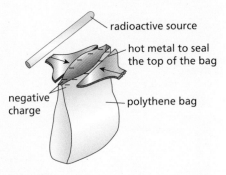

radioactive source

hot metal to seal the top of the bag

negative charge

polythene bag

FIG 5.4.5

Review of Static electricity

- Atoms are the basic building blocks from which all things are made.

- An element is a substance that cannot be made into simpler substances.

- Elements may be metals, non-metals or semi-metals.

- An atom is the smallest particle of an element that can exist.

- An atom contains the sub-atomic particles: protons, neutrons and electrons.

- Protons and neutrons make up the nucleus of an atom, while electrons are found outside the nucleus.

- Protons and neutrons have a relative mass of 1 while electrons have a relative mass of 0.

- Protons and electrons have equal but opposite charges.

- Static electricity is electrical charge built up on an insulator.

- Insulators can be charged by friction.

- Electronic charge is the result of electrons being transferred between insulators.

- Insulators that lose electrons become positively charged, while insulators that gain electrons become negatively charged.

- Like charges repel while unlike charges attract.

- Areas of an insulator may carry opposite charges due to electromagnetic induction.

- Electrostatic charge sometimes makes materials behave in unusual ways.

- Static electricity has important uses including: spray painting and photocopying.

- Static electricity creates some hazards including: damage by lightning strikes, build-up and discharge when liquids and powders pass along pipes, build-up of charge during polythene manufacture.

Review questions on Static electricity

1. Draw a diagram of an atom that contains two protons, two neutrons and two electrons. Choose appropriate symbols for these particles and include a key alongside your drawing.

2. **a)** What does the word 'static' mean?

   b) Why is there sometimes lightning during a storm?

   c) Why does static electricity build up on a car body?

3. Fig 5.RQ.1 shows details of a device that refreshes the air in a room by removing particles of dust.

   Explain how this device works.

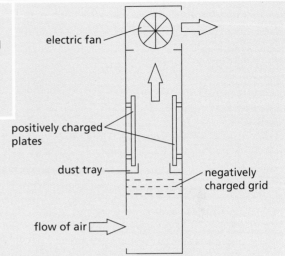

FIG 5.RQ.1

4. Explain each of the following observations.

   a) When a person combs their hair with a plastic comb, the comb becomes charged but this does not happen when they use an aluminium comb.

   b) Electronic components are sometimes supplied in antistatic bags.

   c) Clothing made of synthetic fabrics sometimes crackles when pulled apart after drying in a tumble drier.

   d) A rescue person hanging from a helicopter by a rope carries an electrostatic charge.

   e) The screen of a television becomes dusty even when the rest of the television is clean.

5. Diamonds are electrical insulators. They are found in dirt which conducts electricity. Fig 5.RQ.2 shows an electrostatic method for separating diamonds from dirt.

   a) Explain how the method works.

   b) The particles are closely spaced at W on the conveyor belt. Explain why they are further apart at X.

   c) Explain the curved path of the diamonds between Y and Z.

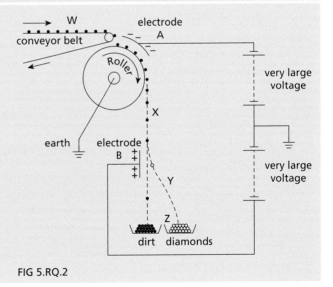

FIG 5.RQ.2

Building an electroscope

A gold-leaf electroscope is a device that scientists use to detect electrostatic charge.

It consists of a metal cap attached to a metal rod. The other end of the metal rod is flattened into a plate and a thin piece of gold leaf is attached to it. The metal parts are insulated from the casing.

Here is what happens when a charged rod is brought close to the metal cap.

When an electrostatically charged object is placed near to (but not touching) the cap the gold leaf moves away from the plate. If the object is removed the gold leaf falls back to its original position.

You are going to make an electroscope to extend your studies of static electricity.

FIG 5.SIP.1 A gold-leaf electroscope

1. You are going to work in a group of 3 or 4 to make an electroscope out of commonly available objects and materials. The tasks are:

   - To review how to charge an object.
   - To examine a gold-leaf electroscope.
   - To make an electroscope using whatever objects and materials are available.
   - To test your electroscope and compare how well it works with a gold-leaf electroscope.
   - To modify the design of your electroscope on the basis of what you learnt from testing it.

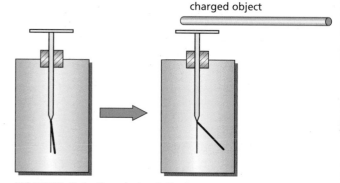

FIG 5.SIP.2 Detecting electrostatic charge

   - To prepare written instructions for how to build a simple electroscope which other students can use to build a similar device. The inclusion of illustrations will be particularly important.

   a) Look back through this unit and make sure you know how to charge an object.

   b) Look carefully at a gold-leaf electroscope or a picture of one. Identify the different parts and take particular note of those which are made of metal and those which are not.

c) Think about how you are going to build an electroscope from everyday objects and materials. For example:

- You won't have any gold foil but would aluminium cooking foil work if it isn't too thick?
- You won't have a metal case with glass sides but would any empty glass jar with a plastic lid work?
- You won't have a metal cap and rod but would a big nail work? How are you going to get the end flat?

The figure shows some examples of home-made electroscopes. They might provide ideas.

Don't forget to take some photographs during construction. These will be useful when you come to write the instructions for you to build your device.

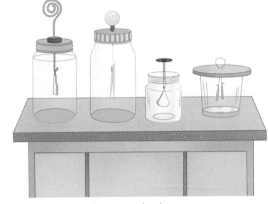

FIG 5.SIP.3 Some home-made electroscopes

d) When you have completed your electroscope try it out with some differently charged objects. Compare how well it performs against a gold-leaf electroscope. You will need to analyse your design. Try asking yourself some questions like:

Is the metal foil too thick to rise effectively?

Where can I get some thinner foil?

Would the electroscope work better if the metal cap was bigger?

How can I make it bigger?

e) Modify the design of your electroscope by incorporating those features that you have identified as a result of testing. Carry out more tests until you are satisfied that your electroscope is as good as it can be.

f) Write a series of instructions to enable another student to build a device like yours. Before you start have a look at some instructions for building other things. For example, have your parents built any self-assembly furniture lately? How were the instructions laid out? How much use was made of illustrations?

You will find that instructions are generally supported by lots of diagrams and other illustrations. People often find these easier to follow than written instructions alone.

When you have completed your instructions you might be asked by your teacher to describe the construction of your electroscope to the class.

Unit 6: Current electricity

We are learning how to:

- distinguish between electrical insulators and conductors
- relate flow of current to conduction.

Conductors and insulators ≫

Electricity is a form of energy. It can pass easily through some materials but not others.

- Materials that allow an electric current to flow through them are called electrical **conductors**.

- Materials that prevent the flow of an electric current are called electrical **insulators**.

Activity 6.1.1

Conductors and insulators

Here is what you will need:

- Battery of three cells
- Lamp
- Two crocodile clips
- Connecting wires
- Samples of materials, e.g. aluminium, copper, plastic, rubber, wood.

Here is what you should do:

1. Connect the components of the circuit together as shown in Fig 6.1.1.

2. Before testing the sample materials, test the circuit by touching the crocodile clips together. If the circuit is complete the lamp should light up. If the lamp does not light up, check all the connections.

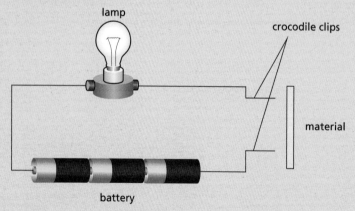

FIG 6.1.1

3. Take the first sample of material and clip the crocodile clips to each end of it. If the lamp lights the material is a conductor. If the bulb does not light the material is an insulator.

4. Present your observations in the form of a table. On one side write the names of the conducting materials, and on the other write the names of the insulating materials.

Metallic structure

Metals are all excellent conductors of electricity. To understand why, we need to consider **metallic structure**. Metals consist of a matrix of particles surrounded by a 'sea' of negatively charged **electrons**. These electrons are delocalised and free to move about.

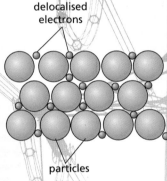

delocalised electrons

particles

FIG 6.1.2 Structure of a metal

When a conductor, such as a metal wire, is connected to a battery, electrons flow through the conductor carrying electrical charge. Electricity is a flow of electrons.

Insulators, such as plastics and glass, do not have delocalised electrons so they are unable to conduct electricity.

Fun fact

The best conductor is the metal silver, followed by copper, then gold and then aluminium. Why are electric wires made of copper and not silver?

Key terms

conductors materials that allow an electric current to flow through them

insulators materials that prevent the flow of an electric current

metallic structure a matrix of particles surrounded by a 'sea' of negatively charged electrons

electrons negatively charged particles

Check your understanding

1. Arrange the following materials into two lists: conductors and insulators.

| aluminium | copper | glass | iron |
| plastic | rubber | steel | wood |

Electricity and safety

We are learning how to:

- use electricity safely.

Electricity and safety >>>

Electricity is potentially dangerous so we must learn how to use it safely.

Some plugs have two pins and some have three. Sockets may have caps to prevent children poking things into them.

FIG 6.2.1 An electrical appliance is connected to the mains electricity supply by a plug, which fits into a socket

Activity 6.2.1

Examining a plug and socket

Here is what you need:

- Plug
- Socket.

 SAFETY

Even though plugs and sockets are made of plastic, they should never be touched with wet hands. There is a danger that wetness will pass into the plug or socket and this can result in an electric shock.

Here is what you should do:

1. Look carefully at the plug and observe where the wires from the appliance will be connected.

2. Notice that when the plug is placed fully in the socket there are no exposed metal parts.

3. Look carefully at the socket and where the wires from the supply will be connected.

4. Notice that the holes for the pins of the plug are very small, which prevents people accidentally pushing something into the holes.

5. Notice also that the plug pins must be fully inside the socket before they connect with the electricity supply. This means that no parts of the plug pins are exposed.

6. From what type of material are the bodies of the plug and socket made?

Two wires are needed to make a circuit. Some plugs have three pins and three wires. The third wire is called the **earth wire**. It protects the user should a fault develop in the appliance.

If the wires in a metal table lamp came loose and touched the metal body, anyone touching it would receive an electric shock. The earth wire is connected to the metal body. Most of the current will pass through it so the user would get a small shock and not come to any harm.

FIG 6.2.2 Metal-bodied appliances should be earthed

FIG 6.2.3 Plastic-bodied appliances do not need to be earthed

An electrical appliance that has a plastic body does not need to be earthed because plastic is an insulator. Such appliances are described as **double insulated**.

Check your understanding

1.

FIG 6.2.4 Gabriella's hairdryer

a) How does Gabriella connect her hairdryer to the power supply?

b) Why should Gabriella not use her hairdryer when her hands are wet?

c) What is the outer casing made of?

d) Will Gabriella's hairdryer have a 2-pin or a 3-pin plug? Explain your answer.

Key terms

earth wire wire in a plug that is connected to the metal body of the appliance to protect the user in the event of a fault

double insulated an appliance that has a body made of an insulating material

Complete circuit

We are learning how to:

- construct simple electrical circuits
- identify a complete circuit.

Complete circuit ›››

We use mains electricity to power most appliances in our homes but it is far too dangerous for building circuits in the laboratory. Instead we use a **cell** or a **battery**. These provide much less electrical energy than the mains supply.

You might use the words 'cell' and 'battery' to mean the same thing in everyday language but in science these terms have particular meanings. A cell is what is often incorrectly called a battery, and a battery is a combination of two or more cells.

A **circuit** is a complete pathway around which an electric current can flow. The pathway must be made of a conductor such as a metal wire.

FIG 6.3.1 This is a single cell

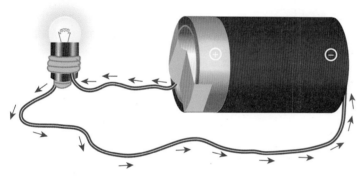

FIG 6.3.3 Direction of conventional current

FIG 6.3.2 When two or more cells are used together to power an electrical device they are called a battery

A cell has a positive (+) terminal and a negative (–) terminal. Conventionally, the direction of the electric current is taken to flow from the positive terminal to the negative terminal. This is called **conventional current flow.**

In reality, the current is the result of a flow of electrons. Since electrons carry a negative charge, they actually flow from the negative terminal (where there are a lot of negative charges) to the positive terminal of the cell (where there are fewer negative charges). This is called **electron flow** and is in the opposite direction to the conventional current.

Activity 6.3.1

Building simple circuits

Here is what you need:

- Cell in holder
- Lamp
- Connecting wires.

Here is what you should do:

1. Make the circuit shown in Fig 6.3.4.

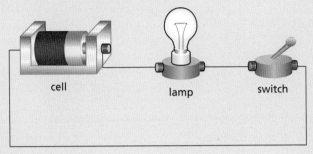

FIG 6.3.4

2. Draw the circuit and, alongside, say whether the lamp lit up or not when the switch was closed.

3. Make some other circuits and make a drawing of each and say whether the lamp lights or not when the switch is closed.

4. Look at the circuits you made in which the lamp lit up. Can you see anything similar about them?

Check your understanding

1.

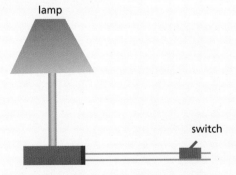

FIG 6.3.6 A lamp connected to the mains electricity supply

Explain, in terms of flow of current, how the switch is able to control the current in the lamp circuit.

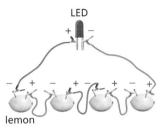
Key terms

cell electrical power source often incorrectly called a battery

battery a combination of two or more cells

circuit a complete pathway around which an electric current can flow

conventional current flow current taken to flow from the positive terminal to the negative terminal of a cell

electron flow electrons flow from the negative terminal to the positive terminal of a cell

Cells and lamps

We are learning how to:

- construct simple electrical circuits
- observe the effect of using different numbers of cells and lamps in a circuit.

Cells and lamps ⟫

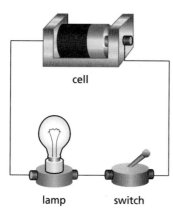

FIG 6.4.1 A simple lighting circuit might consist of a cell, a lamp and a switch

When additional components are added to a circuit, they can often be added in different ways, with different results.

Activity 6.4.1

Making circuits with cells, lamps and switches

Here is what you need:

- Two cells
- Two lamps
- Switch
- Connecting wires.

Here is what you should do:

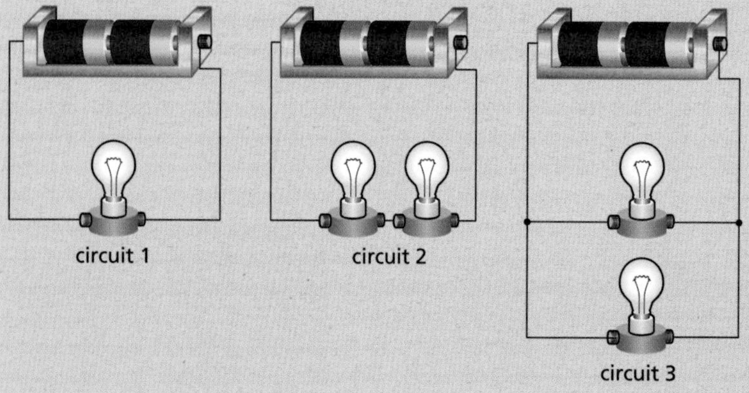

circuit 1   circuit 2   circuit 3

FIG 6.4.2

1. Build circuit 1.

2. Reverse the direction of one of the cells in circuit 1. Does the lamp still light up?

3. Build circuit 2 and make a note of how brightly the lamps glow.

4. Build circuit 3.

5. Do the lamps in circuit 3 glow less brightly, as brightly or more brightly than in circuit 2?

6. Include a switch at some different places in circuit 3. Do the lamps turn on and off differently according to the position of the switch?

There is a difference in potential energy (**potential difference**) between the terminals of a cell. This is measured in **volts (V)**. A single cell has a potential difference of 1.5 V, which is usually written on the side of it.

a)

1.5 V + 1.5 V = 3.0 V

b)

1.5 V – 1.5 V = 0.0 V

FIG 6.4.3 **a)** When two cells are arranged so that their terminals point in the same direction, the overall potential difference of the battery is the sum of the potential differences of the cells, i.e. 3.0 V

b) When cells are arranged so that their terminals point in opposite directions, the overall potential difference of the battery is the difference between the potential differences of the cells, i.e. 0 V

lamps in series

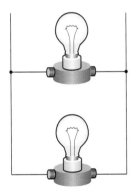

lamps in parallel

FIG 6.4.4 Two lamps can be connected in a circuit in two different ways

A car battery really is a battery in the scientific sense of the word.

It consists of six cells joined together inside a tough rubber casing. The potential difference of each cell is 2.0 V so the potential difference of the battery is 2.0 V added six times, i.e. 6 × 2 = 12.0 V.

Check your understanding

1. Draw a circuit containing two cells, two lamps and one switch so that both lamps are on when the switch is closed and one lamp remains on when the switch is open.

Key terms

potential difference difference in potential energy between two points

volts (V) unit of potential difference

Measuring current

We are learning how to:

- construct simple electrical circuits
- measure current using an ammeter.

Measuring current 》》

The amount of current passing in a circuit is measured using an **ammeter**. The unit of current is the **ampere (amp, symbol A)**.

In the activities we carry out in the laboratory, the current is often less than 1 ampere so we measure current in milliamperes or mA. There are 1000 milliamperes in 1 ampere.

$$1 \text{ A} = 1000 \text{ mA}$$

There are two types of ammeter in common use. An analogue ammeter has a moving pointer and the current is read from a scale where the pointer stops. A digital ammeter gives a direct numerical readout. It may be part of a **multimeter**, which has many different uses.

The positive terminal of the ammeter is connected to the positive terminal of the cell, and the negative terminal is connected to the negative terminal of the cell.

FIG 6.5.1 **a)** Analogue ammeter **b)** Digital ammeter

Activity 6.5.1

Measuring the current at different points in a circuit

Here is what you need:

- Battery containing two cells
- Three lamps
- Ammeter
- Connecting wires
- Long length of nichrome wire
- Crocodile clip.

1. Build the circuit shown in Fig 6.5.3.

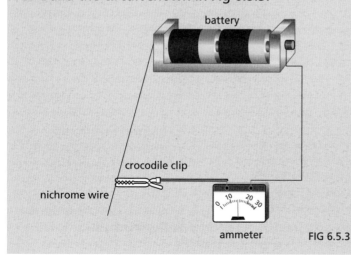

FIG 6.5.3

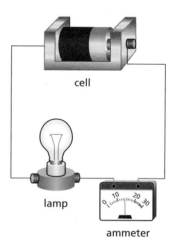

FIG 6.5.2 An ammeter is always connected in series to the component whose current it is measuring

2. Connect the ammeter to different points on the nichrome wire using a crocodile clip.

3. How does the reading on the ammeter change with changes in the length of the nichrome wire in the circuit?

4. Build the circuit shown in Fig 6.5.4.

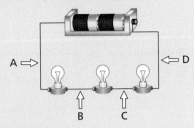

FIG 6.5.4

5. Connect the ammeter at point A in the circuit and record the current.

6. Repeat this at points B, C and D, recording the current each time.

7. What can you say about the current at different points in this circuit?

Nichrome wire resists the flow of current in a circuit. The greater the length of nichrome wire included in a circuit, the smaller the current that flows through it.

When components such as lamps are connected in series in a circuit, the current is the same no matter where in the circuit it is measured.

Check your understanding

1. a) What is the reading on the ammeter:

   i) in milliamperes?

   ii) in amperes?

   b) Draw a diagram to show how an ammeter should be placed in a circuit containing a battery of two cells and one lamp, so that the current flowing through the lamp can be measured.

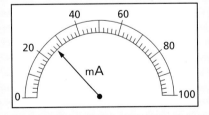

FIG 6.5.5 The reading on a milliammeter

Fun fact

Electrical appliances are rated according to the amount of current that flows through them. Appliances that have small currents include lights, computers and televisions. Appliances that have high currents are those that supply heat in some way such as irons, kettles and ovens.

Key terms

ammeter instrument used to measure current in a circuit

ampere (amp, symbol A) unit of current

multimeter instrument used to measure current and other properties of a circuit

Circuit symbols

We are learning how to:

- represent simple circuits using diagrams
- identify components of an electrical circuit from their symbols.

Circuit symbols

It would be possible to draw all of the **components** in an electrical circuit diagram but this would be time-consuming. It is much easier to draw the circuit using **symbols** to represent each of the components.

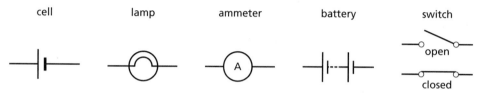

FIG 6.6.1 These symbols for components are used and understood by scientists all over the world

Notice that the symbol for a cell has two vertical lines. The long thin line represents the positive (+) terminal and the shorter thicker line represents the negative (–) terminal. It is important that you draw the symbol in the correct direction in a circuit diagram.

To draw a **circuit diagram** we draw the appropriate symbols and then connect them together by lines to represent connecting wires.

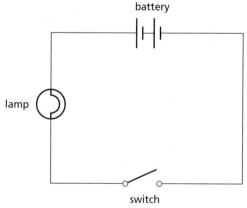

FIG 6.6.2 A circuit diagram for a circuit containing a battery, a lamp and a switch all connected in series

Notice that if a battery is composed of several cells, repeating the symbol for a cell can be tedious. It is much easier to join the symbols for two cells by a dashed line.

FIG 6.6.3 How to represent a battery composed of six cells

Activity 6.6.1

Drawing circuit diagrams

You will not need any equipment or materials for this activity.

Here is what you should do:

1. Look back at Fig 6.5.2 and redraw the circuit using suitable symbols.

2. Look back at Fig 6.5.3 and redraw the circuits using suitable symbols.

3. Look back at Fig 6.5.4 and redraw the circuit using suitable symbols.

4. Would you say that it is easier to draw circuits using symbols than drawing the components?

5. Would you say that circuits drawn in symbols are easier to understand than circuits in which the components are drawn?

Check your understanding

1. Name the components in this electrical circuit.

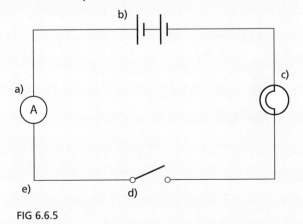

FIG 6.6.5

2. Draw a circuit diagram containing a battery of three cells, two lamps and a switch connected in series.

Fun fact

There are two symbols commonly used to represent a lamp in circuit diagrams.

lamp as a source of light   lamp as an indicator

FIG 6.6.4 Symbols for a lamp

One symbol is used when the lamp is a source of light, such as in a torch circuit. The other is used when the lamp is an indicator of some kind, such as a light that comes on when an appliance is in use.

Key terms

components parts of a circuit

symbols signs used to represent something

circuit diagram diagram showing how components in a circuit are connected

Constructing circuits from circuit diagrams

We are learning how to:

- represent simple circuits using diagrams
- construct a circuit from a circuit diagram.

Constructing circuits from circuit diagrams ⟫

To build a circuit we need to examine a circuit diagram in order to:

- identify the electrical components

- determine how the components are connected together.

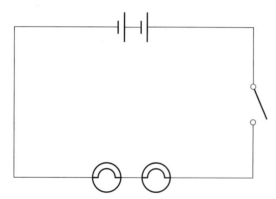

FIG 6.7.1 The symbols show that the circuit contains: two cells, two lamps and one switch connected in series

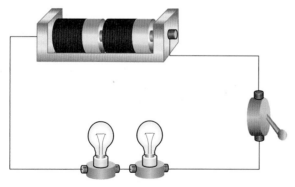

FIG 6.7.2 The information from the circuit diagram allows us to build the actual circuit

Building circuits from circuit diagrams

Here is what you need:

- Three cells
- Two lamps
- Two switches
- Ammeter
- Connecting wires.

Here is what you should do:

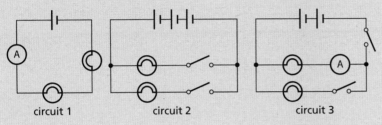

circuit 1   circuit 2   circuit 3

FIG 6.7.3

1. Select the components you will need to build circuit 1 and connect them together as shown in the diagram.

2. Check your circuit with the circuit diagram to make sure they are the same.

3. Repeat this for circuit 2 and then circuit 3.

Check your understanding

1. Fig 6.7.4 is a circuit diagram.

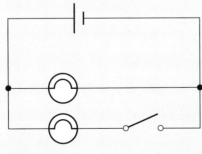

FIG 6.7.4

What components are needed to build this circuit (not including connecting wires)?

Fun fact

Sometimes wires may cross over each other in a circuit but may not actually join together.

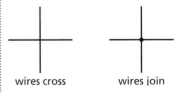

wires cross   wires join

FIG 6.7.5 In order to show the difference between wires that cross and wires that join in a circuit diagram, we place a dot where wires join

Connecting components in series

We are learning how to:

- connect components in series in a circuit.

Connecting components in series ≫

The brightness of a lamp is determined by how much electrical energy is being converted to light (and heat) energy. The brightness is therefore a good indicator of the amount of current passing.

Activity 6.8.1

Investigating bulbs connected in series

Here is what you need:

- Battery containing three cells
- Four lamps
- Ammeter
- Connecting wires.

Here is what you should do:

1. Connect a single lamp in series with an ammeter in a circuit.

2. Note the brightness of the lamp and the reading on the ammeter.

3. Repeat this for two, three and four lamps connected in series and, in addition, find out what happens when one lamp is partially unscrewed from its holder so it goes out.

4. Record your observations in a table.

5. Comment on how the brightness of the lamps changes as the number increases.

6. Comment on how the current changes as the number of lamps increases.

FIG 6.8.1 What happens when increasing numbers of lamps are connected in series

As more lamps are added to the circuit, the lamps shine less brightly. If an ammeter is included in each circuit, it will show that the current falls as the number of lamps increases.

Lamps in **series** are connected by a single circuit. If there is a break in the circuit, such as will occur if one of the lamps burns out or is removed, then the circuit is no longer complete and all the lamps will go out.

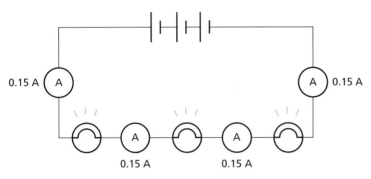

FIG 6.8.2 When lamps are connected in series it does not matter where the ammeter is positioned as the current through the circuit is the same at all points (in this circuit the current is 0.15 A)

Check your understanding

1. Fig 6.8.3 shows a circuit in which three lamps are connected in series. The reading on the ammeter is 0.12 A.

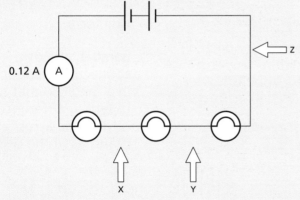

FIG 6.8.3

a) How would the brightness of the lamps change, if at all, if another lamp were added in series?

b) At which of the points X, Y and Z in the circuit would the current be 0.12 A?

c) What would be the reading on the ammeter if one of the lamps burned out? Explain your answer.

Fun fact

When identical cells are connected in series in a battery, the potential difference across the battery is the sum of the potential differences of the cells, provided they are connected in the same direction.

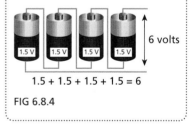

FIG 6.8.4

Key term

series way of connecting components so that they are in one loop in the circuit

Connecting components in parallel

We are learning how to:

- connect components in parallel in a circuit.

Connecting components in parallel ⟫

An electric current only flows through a complete circuit. When two components are connected in **parallel** there are effectively two circuits with a part that is common to both components.

Activity 6.9.1

Investigating lamps connected in parallel

Here is what you need:

- Battery containing three cells
- Four lamps
- Connecting wires.

Here is what you should do:

1. Connect a single lamp in a circuit.

2. Note the brightness of the lamp.

3. Repeat this for two, three and four lamps connected in parallel and, in addition, find out what happens when one lamp is partially unscrewed from its holder so it goes out.

4. Comment on how the brightness of the lamps changes as the number increases.

Key term

parallel way of connecting components in a circuit so that the potential difference is the same across all branches of the circuit

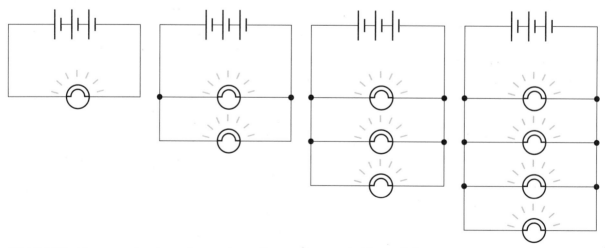

FIG 6.9.1 What happens when increasing numbers of lamps are connected in parallel?

Adding lamps in parallel does not alter their brightness. However, it is not possible to keep on adding more and more lamps without end. There will come a time when the battery is unable to provide sufficient electrical energy.

When two lamps are connected in parallel, they are brighter than they would be if they were connected in series, but they draw twice as much current from the battery. This means that the battery will be exhausted more quickly. Two lamps connected in series will shine less brightly but will continue to shine for longer because the battery will last for longer.

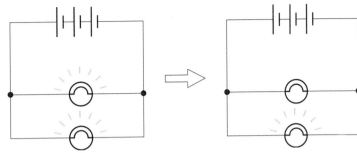

FIG 6.9.2 When lamps are connected in parallel, if one breaks, the circuit containing the second lamp remains complete and so the second lamp remains lit

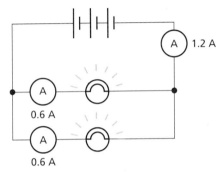

FIG 6.9.3 If ammeters are at different positions in the same circuit, the total current flowing from the battery is equal to the sum of the currents passing through each bulb, so here the values are 0.6 A + 0.6 A = 1.2 A

Check your understanding

1.

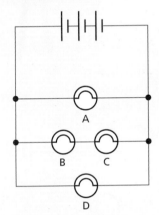

FIG 6.9.4 Four similar lamps connected to a battery

Copy and complete the table to show which lamps would remain on and which would go off when each of the lamps burned out.

Lamp that is burned out	Lamp goes out or remains on			
	A	B	C	D
A	off			
B		off		
C			off	
D				off

TABLE 6.9.1

Fun fact

When identical cells are connected in parallel, the potential difference across them is the same as the potential difference across one cell provided they are connected in the same direction.

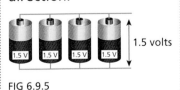

FIG 6.9.5

Voltage and current in a circuit

We are learning how to:

- measure voltage across components in a circuit
- measure current within a circuit.

Voltage >>>

In an electric cell there is a **potential difference** or a difference in energy levels between the two terminals. This potential difference is applied to any circuit to which the cell is connected. It is measured in volts (V) and for that reason is sometimes called the **voltage**.

Measuring voltage >>>

Voltage is measured by a **voltmeter**. As with ammeters, voltmeters may be analogue or digital. A digital voltmeter may be part of a multimeter which can also be used as an ammeter depending on which function is selected.

A voltmeter is connected in parallel to the component to measure the voltage across it. The positive terminal of the voltmeter is connected to the positive side of the component, while the negative terminal of the voltmeter is connected to the negative side.

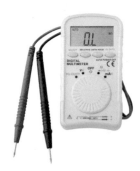

FIG 6.10.1 Voltmeters

Current and voltage >>>

Current and voltage are two different characteristics of the electrical energy passing around a circuit. An electric current only flows around a circuit if there is a voltage difference between the two terminals of the cell or other power source.

The sizes of the current and voltage in a circuit are linked. If you change the voltage applied to a circuit, this will also alter the current that passes.

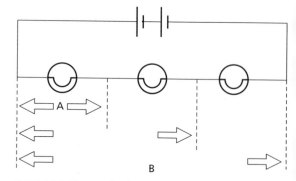

FIG 6.10.2 The total voltage across the three bulbs is equal to the sum of the voltages across each bulb i.e. B = 3A

Activity 6.10.1

Investigating voltage and current in a circuit

Here is what you need:

- A battery containing two cells
- Lamp
- Variable resistor
- Ammeter
- Voltmeter
- Connecting wires.

A variable resistor is a component that resists the flow of current. Altering its value alters the size of its resistance.

Here is what you should do:

1. Connect the components to build the circuit shown in Fig 6.10.3.

2. Copy Table 6.10.1.

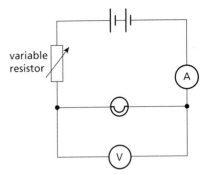

FIG 6.10.3

Voltage across the lamp	Current in the lamp	Brightness of the lamp

TABLE 6.10.1

3. Start by setting the value of the variable resistor to zero and record the voltage, current and brightness of the lamp.

4. Increase the value of the variable resistance to about one quarter, one half, three quarters and all of its value. Each time record the voltage, current and brightness of the lamp.

5. Comment on your observations.

Check your understanding

1. Draw a circuit diagram to show the following:

   • A single lamp connected to a battery containing three cells.

   • An ammeter to measure the current in the lamp.

   • A voltmeter to measure the voltage across the lamp.

2. Four identical lamps were connected in series in a circuit (Fig 6.10.4).

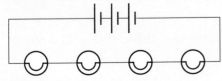

FIG 6.10.4

Comment on the size of the following:

   a) The voltage across each individual lamp.

   b) The total voltage across the lamps.

Key term

potential difference the difference in energy levels at different points in an electric circuit

voltage alternative name for potential difference

voltmeter device for measuring voltage

Review of Current electricity

- Electricity is a convenient form of energy.

- Materials that conduct an electric current are called electrical conductors.

- Materials that do not conduct an electric current are called electrical insulators.

- Metals are good conductors because they contain delocalised electrons. These particles carry electrical charge along the metal. Insulating materials do not contain delocalised electrons.

- Electricity is potentially dangerous so electrical devices must be handled with care.

- Electrical appliances with metal bodies are connected to an earth wire. This provides a pathway for current in the event of wires coming loose and touching the metal body, to protect the user from electric shock. Electrical appliances with plastic bodies do not require an earth wire and are said to be double insulated.

- A cell is a means of providing a small and safe amount of energy. A battery is formed when two or more cells are joined together.

- A cell has a positive (+) terminal and a negative (–) terminal. When cells are placed together to form a battery they must all be pointing in the same direction.

- In order to flow, an electric current must have a complete circuit.

- An electric current can be measured using an ammeter. The unit of current is the ampere, A, or for smaller currents, the milliampere, mA. There are 1000 milliamperes in 1 ampere.

- A circuit diagram is a method of representing a circuit by a group of connected symbols. Each symbol represents a component in the circuit. Circuit diagrams are quick to draw and will be universally understood because scientists around the world use the same symbols.

- Components in a circuit can be connected in series or in parallel.

- When lamps are connected in series the more lamps there are, the dimmer they are and the less current flows in the circuit.

- When lamps are connected in parallel they shine with the brightness of a single lamp up to the point where the cell or battery cannot provide any more electrical energy.

- Lamps connected in parallel are brighter than lamps connected in series but they draw more current from the battery. This means that the battery will be exhausted more quickly. Lamps connected in series will shine less brightly but will continue to shine for longer because the battery will last for longer.

- There is a difference in potential energy or voltage between the terminals of a cell, and this is expressed in volts.

- Voltage is measured in volts using a voltmeter.

- When the voltage across a circuit changes, the current in it will also change.

Review questions on Current electricity

1. Redraw the circuit in Fig 6.RQ.1 as a circuit diagram, using appropriate symbols for the components.

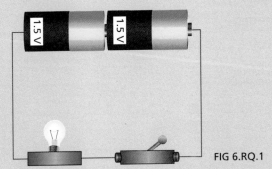

FIG 6.RQ.1

2. Dante used the circuit shown in Fig 6.RQ.2 to test whether materials conduct electricity.

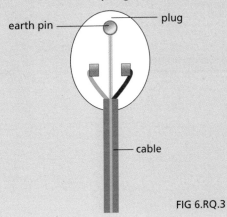

material   FIG 6.RQ.2

a) How was Dante able to decide whether or not each material conducted electricity?

b) Before testing the materials, Dante connected two crocodile clips together. Why did he do this?

c) Here is a list of the materials Dante tested. Arrange them in groups according to whether or not they conduct electricity.

copper   cardboard   plastic   iron
wood   glass   steel   lead

d) What name is given to a material that does not conduct electricity?

e) Give one common feature of all of the materials that conduct electricity.

3. Fig 6.RQ.3 shows the inside of the electric plug fitted to an electric iron.

earth pin ———   ——— plug

——— cable

FIG 6.RQ.3

a) i) How many wires are in the cable?

  ii) Which colour wire goes to the earth pin?

b) The cable of a plastic-bodied reading lamp only has two wires.

  i) Which wire is missing?

  ii) Why is it not needed?

c) Part of the plastic top of the plug has broken off.

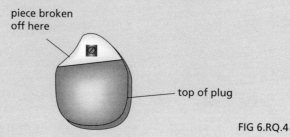

piece broken off here

top of plug

FIG 6.RQ.4

Is the plug still safe to use? Explain your answer.

4. Fig 6.RQ.5 shows a circuit containing a cell, two lamps and an ammeter.

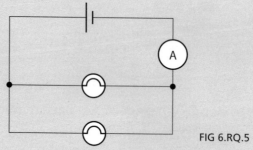

A

FIG 6.RQ.5

State whether each of the following circuits is equivalent to the above circuit or not.

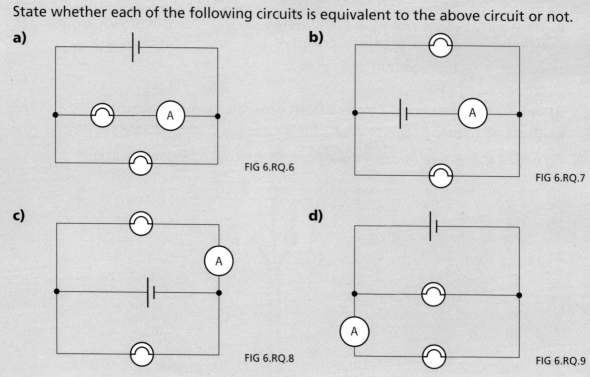

a)

A

FIG 6.RQ.6

b)

A

FIG 6.RQ.7

c)

A

FIG 6.RQ.8

d)

A

FIG 6.RQ.9

5. FIG 6.RQ.10 shows an experiment using some circuits containing identical lamps.

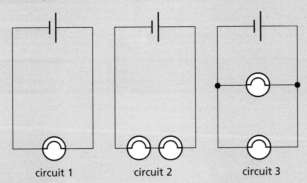

FIG 6.RQ.10

a) If the brightness of the lamp in circuit 1 is 'normal', how bright are the lamps in circuit 2 and circuit 3 compared to circuit 1?

b) If the cells used in these circuits are identical at the start of the experiment, how long will the cells in circuit 2 and circuit 3 last compared to the cell in circuit 1?

c) Explain what will happen if one of the lamps burns out in:

   i) circuit 2

   ii) circuit 3.

6. Shivana made the following circuits. Look carefully at each one and say whether the bulb would come on or not when the switch was closed.

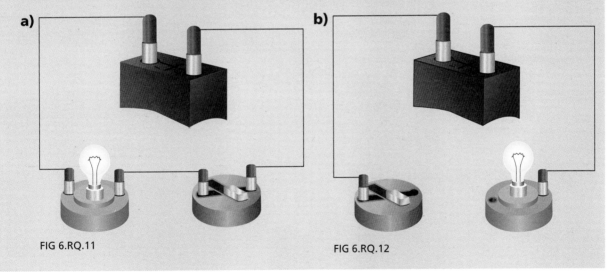

a)

FIG 6.RQ.11

b)

FIG 6.RQ.12

c)

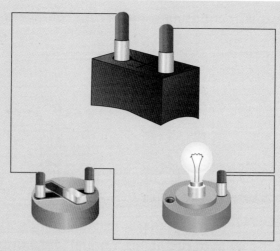

FIG 6.RQ.13

d)

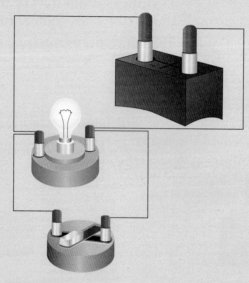

FIG 6.RQ.14

e)

FIG 6.RQ.15

7. Fareed wants to make a circuit containing two cells, a lamp and a switch so that the lamp lights up when the switch is closed, but he is having problems.

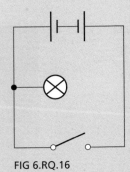

FIG 6.RQ.16

a) There are three problems with Fareed's circuit. Explain what they are.

b) Redraw the circuit showing the components drawn correctly.

8. Fig 6.RQ.17 shows the components inside a toy car.

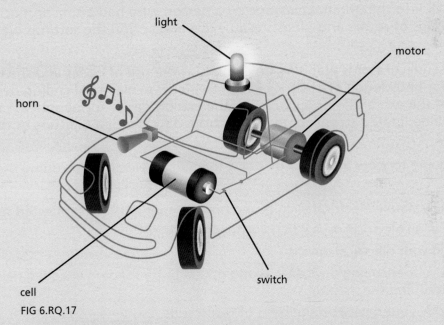

FIG 6.RQ.17

a) Which of the labelled components provides the car with energy?

b) Draw a circuit diagram showing how the components are connected together. Make up suitable symbols for the horn and the motor.

c) If the horn in the toy car broke, would the lamp still work? Explain your answer.

Building an encoding machine

During wars and even peace times, rival countries often want to send messages to their troop commanders or spies which they don't want the opposition to be able to read. To do this they encode their messages.

During the Second World War the Germans had a special device called the enigma machine. It consisted of a keyboard together with a plug board and several rotors. Using different settings of rotors and plugs

FIG 6.SIP.1 An enigma machine

they could send coded messages which were very difficult for the allied intelligence to decipher. The coded message appeared to be a set of random letters and could only be converted back into the original message by a person who had a similar machine with the same settings of rotors and plugs. As an extra precaution the settings were changed every day.

The local museum curator is planning an exhibit on the use of codes and she has asked you to use your knowledge of electric circuits to make an electrical coding device that will be part of the exhibit. In order to make the exhibit easier for people to understand the device will be limited to encoding words formed by the first 6 letters of the alphabet only.

1. You are going to work in groups of 3 or 4 to build an electrical encoding device. The tasks are:

    • To review simple electrical circuits and how components are connected together.

    • To devise an electrical device that encodes a message.

    • To build your electrical device.

    • To test your encoding device by asking others to try sending encoded messages and then decoding the messages.

    • To modify your device on the basis of test results.

    • To make your device the centrepiece of an exhibit in which observers are told how the device works and how messages are encoded and decoded.

    a) Look back through the unit and be sure that you recognise the need for a complete circuit to be in place in order for an electric current to flow.

    b) The enigma machine is a very complicated device. You need to think about how you will show encoding in a much simpler way. FIG 6.SIP.2 might give you some ideas. Don't be put off by all of the wires. The two symbols you may not have seen before represent light-emitting diodes (LEDs). Which we often use in place of lamps because they require less current, and a resistor which limits the amount of current that can pass through the LEDs. Your teacher will explain more about LEDs and provide you with a suitably-sized resistor.

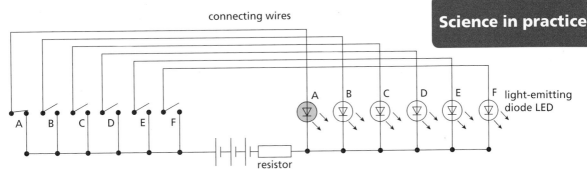

FIG 6.SIP.2 Each letter has a light

There are 6 parallel circuits. Each circuit has a switch and an LED. When the switch for A is closed the LED representing A lights up. However look what happens, in FIG 6.SIP.3, when we jumble up the connecting wires.

Now when the switch for A is closed, the LED representing D lights up. If we were to send each of the letters in the word 'FACED' through our encoder we would get 'EDFAB'!

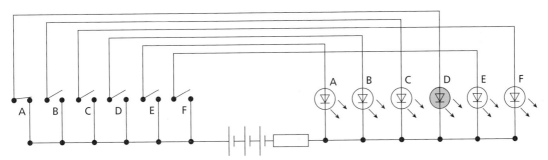

FIG 6.SIP.3 Now the letters are encoded

Can you see how the encoded word might be decoded using a similar set of circuits?

Exhibits that you can look at are interesting but interactive 'hands on' exhibits which you can touch and play with are more fun. In designing your exhibit can you see a way of making it hands on?

c) Once you have designed your device you need to assemble the components you need to build it. Check each part of any circuit that you build as you go along rather than leaving it until the end. You should ask your teacher for help if you are uncertain how to convert your circuit diagram into an actual circuit.

d) In the museum, each exhibit has information about it, together with instructions, if it is interactive. You need to think about these and compile them before testing since they will be an important part of your final exhibit.

Ask a range of people of different ages and interests to try out your device. Make sure you provide sufficient detail so that people who know little or nothing about electrical circuits can appreciate how it works.

e) Once you have completed testing your exhibit you should make whatever changes you think are appropriate to your circuitry and to your information and instructions.

Your work should be as attractive as possible so that people are drawn to your exhibit. Use different sized letters and colours to catch peoples' attention so that they will want to see what your exhibit is all about.

Unit 7: Magnetism

We are learning how to:

- demonstrate the effects of magnetic forces
- determine whether a material is magnetic or not.

Magnetic and non-magnetic materials »

Lodestone is an oxide of iron and is also called magnetite.

FIG 7.1.1 Lodestone, a type of rock, attracts objects made of iron, such as nails and bolts

Activity 7.1.1

Magnetic and non-magnetic materials

Here is what you need:

- Magnet
- Objects made of different materials, e.g. nail, paper clip, plastic ruler, eraser.

Here is what you should do:

1. Place one end of the magnet near an object and find out if the material is attracted to it.

2. Materials that are attracted by a magnet are described as magnetic materials.

3. Test each object in turn with the magnet.

4. Display your observations in a table. On one side of the table list the magnetic materials and on the other side list the non-magnetic materials.

Materials such as iron and steel, that are attracted to a magnet, are described as **magnetic**, while materials such as brass, copper and aluminium, that are not attracted to a magnet, are described as **non-magnetic**.

Permanent magnets

Materials that keep their magnetism for a long time are called permanent magnets.

Materials that have permanent magnetism are iron, mild steel, cobalt and nickel. Modern magnets are often made of special alloys containing these metals, such as alnico and alcomax.

Ceramic or ferrite magnets are made by baking iron oxide and other metal oxides in a ceramic matrix.

FIG 7.1.2 Ceramic magnets can be made in any shape but have the disadvantage that they are brittle, so if they are dropped on a hard surface they will break into pieces

Fun fact

An alloy is a mixture of a metallic element with one or more other elements that may be metals or non-metals. Steel is an alloy of iron and carbon.

Sometimes alloys have more useful properties than the elements from which they are formed. For example, alnico is an alloy of aluminium, nickel and cobalt. It makes more powerful magnets than the pure metals.

FIG 7.1.3 The element neodymium forms alloys with iron and boron that are used to make powerful permanent magnets

Check your understanding

1. Arrange the following metals into two groups: those that are magnetic and those that are not magnetic.

cobalt	copper	gold	iron
magnesium	nickel	steel	zinc

Key terms

magnetic materials such as iron and steel are substances that are attracted to a magnet

non-magnetic materials such as brass, copper and aluminium are substances that are not attracted to a magnet

Law of magnetic poles

We are learning how to:

- demonstrate the effects of magnetic forces
- predict whether two magnetic poles will attract or repel each other.

Law of magnetic poles 〉〉

A magnet has two **poles**: a north pole and a south pole. The north and south poles are usually represented by the letters 'N' and 'S'.

Forces exist between magnets and are concentrated at the poles. The interaction between two magnets depends on the nature of the poles that are brought together.

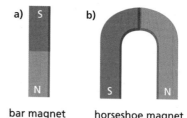

bar magnet   horseshoe magnet

FIG 7.2.1 **a)** Bar magnets are commonly used in the laboratory **b)** A horseshoe magnet is simply a bar magnet that has been bent into the shape of a horseshoe

Activity 7.2.1

Law of magnetic poles

Here is what you need:

- Two cotton loops
- Two bar magnets
- Pencil
- Heavy book.

Here is what you should do:

1. Place a heavy book on top of a pencil so that the pencil is sticking out from the table.

2. Suspend a bar magnet from the pencil using loops of cotton so that it can turn freely.

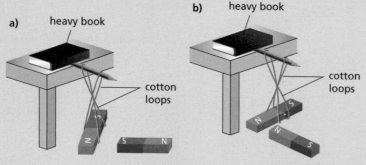

FIG 7.2.2 **a)** Unlike poles **b)** Like poles

3. Bring the N pole of the second magnet towards the N pole of the suspended magnet and record what happens.

4. Bring the N pole of the second magnet towards the S pole of the suspended magnet and record what happens.

5. Repeat steps 3 and 4 but using the S pole of the second magnet.

6. What deductions are you able to make about magnets from your observations?

If one magnet is suspended so it is free to rotate and a second magnet is brought near it:

- if they are **unlike poles**, that is N and S or S and N, the magnets will attract (move towards each other)

- if they are **like poles**, that is N and N or S and S, the magnets will repel (move away from each other).

Check your understanding

1. A compass needle is a magnet. The north pole of the compass always points towards the Earth's magnetic north pole, and the south pole of the compass points towards the Earth's magnetic south pole.

What is the polarity of each of the Earth's magnetic poles? Explain your answer.

FIG 7.2.3 Compass

Fun fact

It is impossible to say if an iron bar is magnetic or not on the basis of whether it is attracted by a magnet.

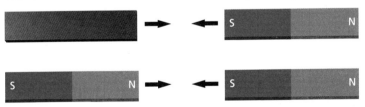

FIG 7.2.4 A magnet would attract an iron bar even if the iron bar were not itself a magnet

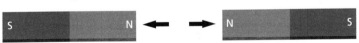

FIG 7.2.5 Repulsion proves an iron bar is a magnet

To test whether an iron bar is a magnet, both ends must be placed near the same magnetic pole of a magnet. If one end is repelled, this proves that the bar is a magnet.

Key terms

pole the end of the magnet

unlike poles two poles that are different, for example, north and south

like poles two poles that are the same, for example, north and north

Magnetic fields

We are learning how to:

- demonstrate the effects of magnetic forces
- draw the magnetic field around a bar magnet.

Magnetic fields >>>

A magnet is surrounded by a pattern of invisible **magnetic field lines**. We can investigate the nature of the field lines using a small plotting compass.

Activity 7.3.1

Magnetic field around a bar magnet

Here is what you need:

- Bar magnet
- Plotting compass
- Plain paper.

Here is what you should do:

1. Place a bar magnet at the centre of a piece of plain paper and draw around it.
2. Remove the magnet and mark the N and S poles on the outline.
3. Place the magnet back on the outline.
4. Place the plotting compass near the north pole of the magnet. Mark two dots on the paper corresponding to the ends of the plotting compass needle.
5. Move the compass a little bit away towards the south pole and repeat drawing the dots.
6. Repeat this procedure until a continuous line is formed around the magnet.

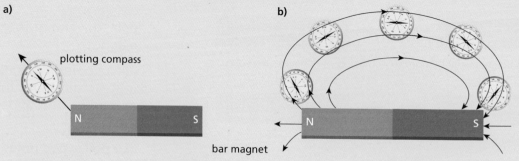

a)

plotting compass

N   S

bar magnet

b)

N   S

FIG 7.3.1

7. Connect all the dots with a smooth curve. This curve is one magnetic field line. Try to obtain four curves on each side of the magnet.

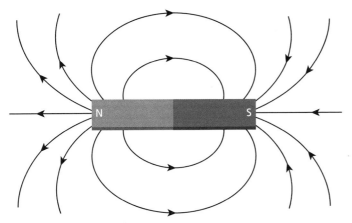

FIG 7.3.2 The magnetic field around a magnet can be represented by a set of magnetic field lines

Magnetic field lines are conventionally shown moving away from a north pole and towards a south pole and arrows are placed on the lines to show the direction of the field. When drawing or interpreting the magnetic field around a magnet you should remember that:

- magnetic field lines never cross over each other
- the **magnetic field strength** is shown by the concentration of field lines and is strongest where the field lines are most dense (at the poles)
- the magnetic force of a magnet decreases with distance from the poles.

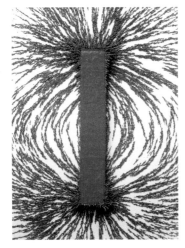

FIG 7.3.4 The magnetic field lines around this bar magnet are also shown by the iron filings around the magnet

Check your understanding

1. Use what you have learned about the magnetic field lines around a bar magnet to draw the magnetic field between the two poles of a horseshoe magnet.

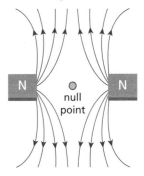
Key terms

magnetic field lines a pattern of invisible lines that shows how the strength and direction of the magnetic field varies around a magnet

magnetic field strength how strong the magnetic field is at a particular point

Magnetic effect of an electric current

We are learning how to:

- describe the magnetic effect of current
- make an electromagnet.

Magnetic effect of an electric current ≫

When a compass needle is placed close to a wire, and then a current is passed through the wire, the compass needle is deflected.

This is called the **magnetic effect of a current**.

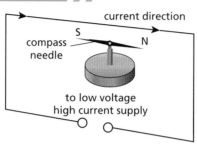

FIG 7.4.1 The magnetic effect of an electric current

Activity 7.4.1

Magnetic field around a wire carrying a current

Here is what you need:

- Thick resistance wire
- Plotting compass
- Plain card
- DC power source.

Here is what you should do:

1. Make a small hole in the middle of a piece of card and push the wire through it.

2. Connect the wire to a DC power source. This has a positive (+) terminal and a negative (−) terminal.

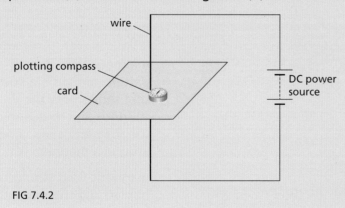

FIG 7.4.2

3. Place a plotting compass near the wire and show the direction that the compass points in by drawing an arrow.

4. Repeat this, placing the plotting compass in different positions until you have built up a map of the field lines around the wire.

5. Reverse the direction of the current through the wire by connecting the wire to the opposite terminals of the power source.

6. Observe if this affects the shape of the magnetic field lines around the wire and the direction of the magnetic field.

Passing a current through a conductor such as a wire creates a magnetic field consisting of a series of concentric circles. The circles are closer together nearer the wire where the magnetic field is strongest.

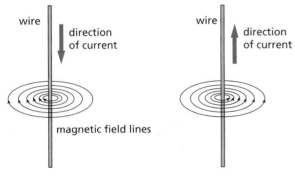

FIG 7.4.3 Magnetic field around a wire carrying an electric current

Reversing the direction of the current in the wire does not alter the shape of the magnetic field, but it does alter the direction of the field lines.

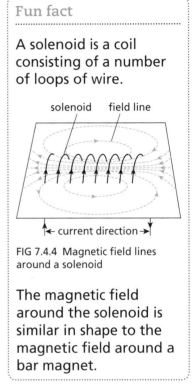

FIG 7.4.4 Magnetic field lines around a solenoid

The magnetic field around the solenoid is similar in shape to the magnetic field around a bar magnet.

Key term

magnetic effect of a current a compass needle is deflected when it is placed close to a wire carrying an electric current

Check your understanding

1. Draw a diagram showing the magnetic field around a wire carrying a current viewed as if you were looking along the wire.

Making an electromagnet

We are learning how to:

- describe the magnetic effect of current
- make an electromagnet.

Making an electromagnet

To make an **electromagnet** of any useful strength, you need to combine the magnetic field around many turns of wire by making a coil or solenoid.

The coils of wire on their own are magnetic. However, if they are wrapped around a steel nail they make an even stronger magnet.

Activity 7.5.1

Making an electromagnet

Here is what you need:

- Steel nail
- Plastic-coated wire
- DC power source
- Paper clips
- Plotting compass.

Here is what you should do:

1. Take a length of wire that is coated in plastic insulation and coil it around a steel nail.

2. Make between 15 and 20 coils of wire around the nail, depending on the length of the wire.

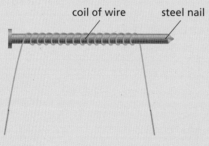

coil of wire   steel nail

FIG 7.5.1

3. Connect your coil to a DC power source.

4. Check that you have made an electromagnet by seeing if metal paper clips are attracted to it.

5. Place a plotting compass at different points around your electromagnet and use the direction the compass points to each time to draw a diagram of the magnetic field lines around your electromagnet.

Electromagnetism is sometimes described as temporary magnetism. An electromagnet is only magnetic while a current flows through it. If the current is turned off, the electromagnet ceases to be magnetic.

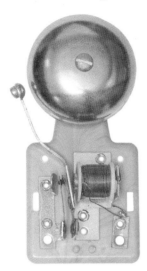

FIG 7.5.2 A practical electromagnet

Electromagnets used in devices such as electric bells consist of many coils of thin copper wire. At first glance the wire might not appear to be insulated, but it is. The wire is covered in a layer of lacquer, which is far less bulky than a plastic coating.

Check your understanding

1. Explain why a coil of wire can only attract, for example, paper clips when an electric current is passing through it.

2. In a scrapyard an electromagnet is used to move metals.

    a) Which metals can be moved by the electromagnet?

    b) Describe how the operator uses the electromagnet to move these metals.

    c) Suggest what the electromagnet looks like inside if the outer casing is removed.

electric cable

electromagnet

FIG 7.5.4

    d) Explain why a permanent magnet would not be suitable for this job.

Key term

electromagnet a magnet produced when a current is passed through a wire or coil of wire

Strength of an electromagnet

We are learning how to:

- describe the magnetic effect of current
- compare the strengths of different electromagnets.

Strength of an electromagnet ⟫

An electromagnet is a coil of wire through which an electric current is passed. What determines the strength of an electromagnet?

- Would wrapping the coil around a piece of wooden dowel be just as good as wrapping it around a steel nail?

- Does it matter how many turns of wire are in the coil?

- Does it matter how much current you pass through the coil?

Activity 7.6.1

Investigating the strength of electromagnets

Here is what you need:

- Steel nail
- Wooden rod
- Plastic-coated wire
- DC power source
- Paper clips.

Here is what you should do:

1. Take a length of wire that is coated in plastic insulation and coil it around a steel nail.

2. Make 20 coils of wire around the nail.

3. Connect your electromagnet to a DC power supply and count how many paper clips it will lift off the desk.

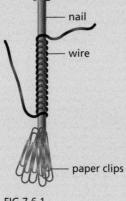

FIG 7.6.1

4. Repeat steps 1 to 3 but use a piece of wooden rod in place of a steel nail.

5. Now make 10 coils of wire around a steel nail.

6. Connect your electromagnet to a DC power supply and count how many paper clips it will lift off the desk.

7. Repeat steps 2 to 6, but this time only use half of the current used previously.

8. From your observations, deduce what factors determine the strength of an electromagnet.

The strength of an electromagnet is increased by:

- wrapping the coils of wire around a core of magnetic metal such as iron or steel

- increasing the number of turns of wire in the coil

- increasing the amount of current passing through the coil.

Check your understanding

1. Say whether each of the following statements is true or false.

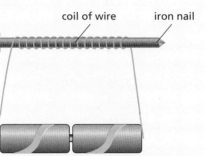

coil of wire   iron nail

FIG 7.6.2 An electromagnet made by wrapping wire around an iron nail and connecting it to a battery

a) Reversing the battery will reduce the strength of the electromagnet.

b) The iron nail would still be a strong magnet even if the battery were removed.

c) Decreasing the number of turns of wire on the coil would reduce the strength of the electromagnet.

d) The electromagnet would be stronger if the iron nail were removed.

e) Wrapping the wire around two iron nails would make the electromagnet twice as strong.

f) Connecting the coil to a battery with a higher voltage would make the electromagnet stronger.

Fun fact

Credit and debit cards have a magnetic strip containing information about the cardholder's account.

FIG 7.6.3 If a card with a magnetic strip is placed too near to another magnet, the magnetic field will corrupt the information and the card will no longer be of any use

Uses of permanent magnets and electromagnets

We are learning how to:

- describe the magnetic effect of current
- explain the uses of permanent magnets and electromagnets.

Uses of permanent magnets and electromagnets ⟩⟩

Permanent magnets and electromagnets have many applications.

Electric bell

Activity 7.7.1

Investigating how an electric bell works

Here is what you need:

- Electric bell
- Power source
- Screwdriver.

Here is what you should do:

1. Remove the cover from the outside of the bell so you can examine the parts.

2. Identify the electromagnet.

3. Turn the bell on and off and observe the effect this has on the components.

4. Use your knowledge of electric circuits and magnets to explain how the bell works.

When the bell switch is pushed the circuit is complete and the following happens:

- The electromagnet becomes magnetic.

- The electromagnet attracts the soft iron armature and the hammer strikes the gong.

- As the soft iron **armature** moves, the circuit is broken and the electromagnet loses its magnetism.

- The springy metal strip moves the armature back to its starting position and the cycle repeats for as long as the switch is pushed.

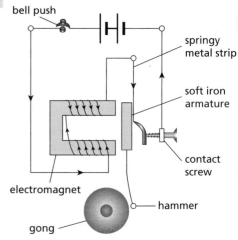

FIG 7.7.1 The structure of an electric doorbell

Relay

A **relay** is a switch that is operated by an electromagnet.

When a current passes between A and B, the soft iron core becomes an electromagnet. The iron armature is attracted to the electromagnet and turns on the pivot. The springy metal contacts at C are closed, completing the circuit connected across D and E.

A relay allows a circuit carrying a large current to be controlled by a second circuit carrying a small current. For example, a large current is needed to start a car engine. It is activated by the ignition switch in the car, which only carries a small current. This means that only thin wires are needed for the ignition circuit.

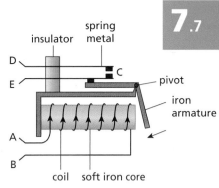

FIG 7.7.2 A relay allows one circuit to be controlled by another

Circuit breaker

A **circuit breaker** is a device that prevents the flow of current in a circuit in the event of a malfunction or fault.

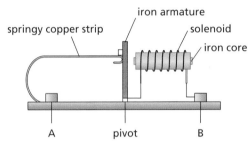

FIG 7.7.3 A simple circuit breaker connected to a circuit at points A and B

In the event of a fault in the circuit, the current passing through the solenoid increases. This increases the strength of the magnetic field around the **solenoid** enough for it to pull the iron armature towards it. When the iron armature moves towards the solenoid, the springy copper strip is released and the circuit is broken.

Check your understanding

1. Explain how an electromagnet is used in an electric bell.

FIG 7.7.4

Fun fact

Circuit breakers are frequently used in place of fuses to protect domestic mains electricity circuits. They work much faster than fuses and can be reset simply by pushing a switch once the fault has been found and rectified.

Key terms

armature a metal part that can move to open or close a circuit

relay a switch that is operated by an electromagnet

circuit breaker a device that prevents the flow of current in a circuit in the event of a fault

solenoid a cylindrical wire coil with a soft iron core inside

Induced current

We are learning how to:

- explain how an electric current may be induced by a magnetic field.

Induced current

Not only does an electric current create a magnetic field, but when a conductor such as a wire is moved through a magnetic field, an **electromotive force**, or e.m.f., is induced in the wire. If the wire forms part of a circuit, an electric current passes in the circuit.

A galvanometer is a very sensitive ammeter that can detect very small currents.

In Fig 7.8.2 a wire has been connected to a galvanometer and held so that a straight section is exactly between the poles of a C-shaped magnet.

Moving the wire in different directions produces different responses on the galvanometer.

FIG 7.8.1 Centre-zero galvanometer

Direction	Response on the galvanometer
1 and 2 (up and down)	The galvanometer deflects in one direction and then in the opposite direction
3 and 4	None
5 and 6	None
Wire stationary	None

TABLE 7.8.1

A current is only induced in a conductor when it cuts across magnetic field lines. If a conductor is held still within a magnetic field, or moves parallel to magnetic field lines, then no current is induced in it.

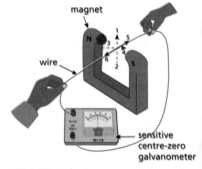

FIG 7.8.2 Inducing a current in a wire

Activity 7.8.1

Inducing a current in a coil

Your teacher will demonstrate this activity.

Here is what you need:

- A centre-zero galvanometer
- A bar magnet
- A coil of wire.

1. Connect the coil to the galvanometer.

2. Push the magnet into the coil, hold it stationary and then pull it out. Observe what happens to the galvanometer while you do this.

3. Experiment by placing the magnet in the coil and then moving it up and down, and side to side. Observe the galvanometer while you do this.

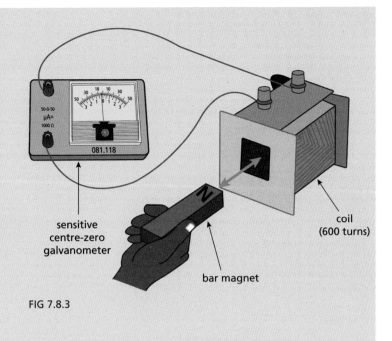

sensitive centre-zero galvanometer

bar magnet

coil (600 turns)

FIG 7.8.3

4. Experiment by varying the speed with which you put the magnet into and out of the coil. Observe the galvanometer while you do this.

The size of the current induced in the coil increases with an increase in the following:

- The speed with which the magnet is moved in and out of the coil.

- The number of turns of wire on the coil.

- The strength of the magnet used.

All of these factors relate to the rate with which the wire cuts the magnetic field lines.

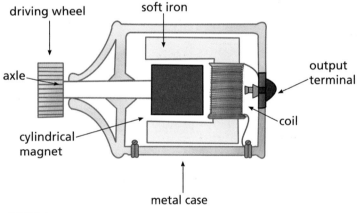

driving wheel

soft iron

axle

cylindrical magnet

metal case

output terminal

coil

FIG 7.8.4

Check your understanding

1. Fig 7.8.4 shows a bicycle dynamo. The cylindrical magnet rotates when the driving wheel rubs against the moving cycle wheel. It is used to provide an electric current to power cycle lights.

   a) Explain how the dynamo works.

   b) Explain why a light connected to the dynamo glows more brightly when the cycle moves faster.

   c) Predict and explain what will happen to the size of the induced current when the cycle stops moving.

Review of Magnetism

- Lodestone or magnetite is a mineral composed of iron oxide that attracts objects made of iron. It is a naturally occurring magnet.

- Materials that are attracted to a magnet are described as magnetic, while other materials are non-magnetic.

- Materials that keep their magnetism for a long time are called permanent magnets.

- The law of magnetic poles states: Like poles repel, i.e. N and N or S and S, and unlike poles attract, i.e. N and S or S and N.

- A magnet is surrounded by a pattern of invisible magnetic field lines.

  o Magnetic field lines are conventionally shown to move away from a north pole and towards a south pole.

  o Magnetic field lines never cross over each other.

  o The strength of the magnetic field is shown by the concentration of field lines and is strongest where the field lines are most dense.

  o The magnetic force of a magnet decreases with distance from the poles.

- When a current passes through a wire, a magnetic field is generated around the wire.

- The magnetic field around a wire carrying a current consists of a series of concentric circles which gradually increase in distance between each other.

- Reversing the direction of the current through a wire does not alter the shape of the magnetic field around it but reverses the direction of the magnetic field lines.

- An electromagnet consists of many turns of wire making a coil or solenoid.

- The first electromagnet was made by the scientist William Sturgeon in 1824. It consisted of about 18 turns of varnished wire wrapped around a piece of iron in the shape of a horseshoe.

- Electromagnetism is sometimes described as temporary magnetism because an electromagnet is only magnetic while a current flows through it. If the current is turned off, the electromagnetic ceases to be magnetic.

- The strength of an electromagnet is increased by:

  o Wrapping the coils of wire around a core of a magnetic metal like iron or steel.

  o Increasing the number of turns of wire in the coil.

  o Increasing the amount of current passing through the coil.

- An electric bell, a relay and a circuit breaker are all common devices that contain an electromagnet.

- An electromagnet can be used to separate iron and steel from other metals in a scrap yard or recycleing plant.

- A current is induced in a conductor, such as a wire, when the conductor is moved in such a way as to cut through magnetic field lines

Review questions on Magnetism

1.

FIG 7.RQ.1 A bar magnet

Copy Fig 7.RQ.1 and draw the magnetic field lines around it. Show the direction of these field lines.

2. A bar magnet was broken into two pieces.

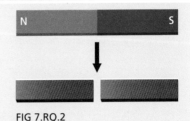

FIG 7.RQ.2

  a) Copy the lower part of the diagram and show the polarity of the new ends formed.

  b) Without using any other apparatus or materials, explain how you could show that both parts of the broken bar magnet have themselves become magnets.

3. Johanna was given four magnets. Her task was to compare how strong they were by counting how many small nails each could lift. Her results are shown in the table.

Magnet	Number of nails
bar magnet	3
C-shaped magnet	6
electromagnet	4
horseshoe magnet	7

TABLE 7.RQ.1

  a) Which magnet was the strongest?

  b) i) Which magnet could be described as a temporary magnet?

    ii) Explain why.

4. Say whether each of the following will make the electromagnet stronger, weaker or have no effect.

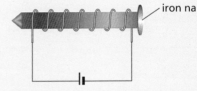

FIG 7.RQ.3 An electromagnet

  a) Increasing the number of turns of wire

  b) Decreasing the current passing through the wire

  c) Putting the nail into the coil from the opposite direction

5. Here are some statements about magnetic field lines. State whether each statement is true or false.

  a) The closer together they are, the stronger the magnetic field.

  b) They sometimes cross over each other.

  c) They come away from a north pole and towards a south pole.

  d) They are always straight lines.

  e) They are not produced by electromagnets.

6. A student is given three iron bars. The ends of the bars are marked A to F.

FIG 7.RQ.4

Two of the bars are known to be magnets and the third one is not. Explain how the student can identify which bar is not a magnet using only the three bars.

7. Fig 7.RQ.5 shows an electromagnet in a circuit with three bulbs: X, Y and Z.

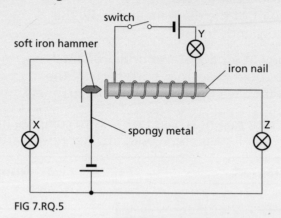

FIG 7.RQ.5

a) Copy and complete the table by indicating whether each bulb is 'off' or 'on' when the switch is open and when it is closed.

Switch	Bulb X	Bulb Y	Bulb Z
open			
closed			

b) The iron nail in the circuit is replaced by a length of wooden dowel of the same diameter.

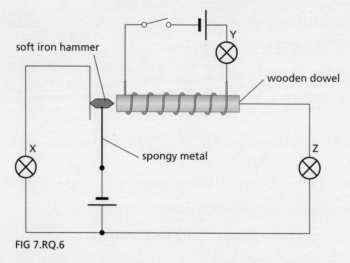

FIG 7.RQ.6

State whether each of the three bulbs will be on or off when the switch is closed. Explain your answer.

8. An investigation was carried out in which bars of three different metals, aluminium, copper and magnetised iron, were hung by a nylon thread so they could rotate.

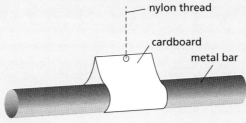

FIG 7.RQ.7

The bar was spun and then allowed to rest. Its position was then recorded. This was repeated ten times for each bar. The results are shown below.

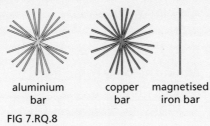

aluminium   copper magnetised
bar   bar   iron bar

FIG 7.RQ.8

a) Describe the pattern shown by the results.

b) Explain the results.

c) Would the result for iron have been the same if an iron bar that was not magnetised had been used? Explain your answer.

9. The diagram below shows a wire being moved vertically upwards between the poles of a C-shaped magnet. A current is induced in the wire.

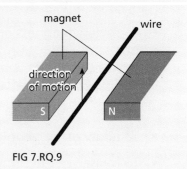

FIG 7.RQ.9

a) The wire is subsequently moved vertically downwards but more slowly than it was moved upwards. State what effect this will have on the induced current.

b) Explain why no current is induced when the wire is moved horizontally between the poles of the magnet.

10. The following diagram shows a reed switch.

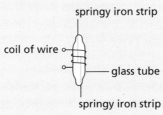

FIG 7.RQ.10

a) Explain why the springy metal reeds bend and touch each other when a current passes through the coil of wire.

b) The springy iron strips can be connected in a circuit. Suggest a suitable application for a reed switch.

Creating a game or an executive toy using magnets

Magnetic football is a game in which small plastic footballers each have a magnet attached to their base. Each player has a stick with a magnet fixed at one end (Fig 7.SIP.1).

When a player moves his/her stick under the table beneath one of their players, the player moves because the magnets attract each other.

An executive toy is a toy that adults play with, and is generally found on their desk at work.

The base of the toy in Fig 7.SIP.2 is a powerful magnet and the bars are made of mild steel. When the bars come into contact with the base, either directly or indirectly, they become small magnets. They 'stick' together and can form lots of different shapes.

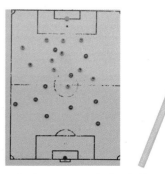

FIG 7.SIP.1 Magnetic football and stick with magnet on the end

A games company has asked you to use your knowledge of magnetism together with your imagination to invent a new game or a new executive toy using magnets.

1. You are going to work in groups of 3 or 4 to make a new game or executive toy. The tasks are:

    *   To review magnets and how they interact with each other. You may wish to use an electromagnet in your device so be sure you understand how to make one and what factors affect its strength.

    *   To make your device.

FIG 7.SIP.2 Executive toy

    *   To allow other people to test your device and comment on it.

    *   To revise your design, making improvements as you think appropriate.

    *   To compile a report for the company in which you describe how the game or toy is made and how it works. Your report should be illustrated by pictures showing the game or toy being used.

    a) Look back through the unit and make sure you understand the laws of magnetic poles. You should also understand how to make an electromagnet and what determines its power.

    b) What sort of game or toy are you going to make? Have a look in catalogues, shop windows, books and the Internet to give yourself ideas.

FIG 7.SIP.3 Magnetic letters

Your game doesn't have to be very sophisticated. For example you might make a simple spelling game by attaching small magnets to the backs of cardboard letters. This is the sort of game that young children might use to learn to spell. The letters need to be attractive.

to switch and battery

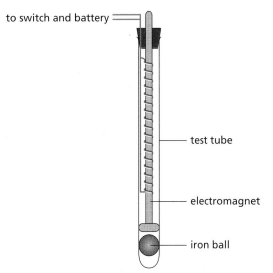

test tube

electromagnet

iron ball

FIG 7.SIP.5 Toy that uses an electromagnet and gravity

FIG 7.SIP.4 Tiny neodymium magnets

Tiny neodymium magnets are very powerful and often used in executive toys. Your teacher will help you to obtain some small magnets.

Your executive toy doesn't need to be overly complicated. For example you might place an electromagnet in a test tube which has an iron ball in the bottom (FIG 7.SIP.5). When the current is switched on the ball is pulled up to the electromagnet and when it is switched off the ball falls.

c) Once your game or toy is made you need to test it. To do this you should pick the sort of people who are going to use it. For example, if you have built a game for young children then you should test it out on young children and observe how they use it.

If you have built an executive toy you should ask some adults to test it and report back to you on how it performed.

Take some pictures of the game or toy being used. These will be a useful addition to your report.

d) As a result of testing, did you observe any problems or did users come back to you with any criticisms or ideas on how improvements might be made?

Don't be put off by criticism. It is useful to have a fresh pair of eyes look at what you are working on.

Modify your game or toy on the basis of observations and comments made during testing. If necessary, run some more testing until you are satisfied what you have made is as good as it can be.

e) Prepare a report which will be sent to the games company along with your game or toy. You should explain how it is used and why you think it will be useful or popular.

Your report should be accompanied by your game or toy and it would be a good thing to include pictures of people using it.

Unit 8: Chemical bonding, formulae and equations

We are learning how to:

- write chemical formulae for elements and for compounds.

A **molecule** is composed of two or more atoms chemically bonded together. A molecule of an element contains only one kind of atom. A molecule of a compound contains atoms of two or more elements.

Chemical formula of elements

Not all elements form molecules. Metallic elements are represented simply by their atomic symbols even though they do not exist as separate atoms, for example Na = sodium, Al = aluminium and Ca = calcium.

Many non-metallic elements exist as molecules composed of two or sometimes more atoms. We show this using subscript numbers after their atomic symbols, for example H_2 = hydrogen, O_2 = oxygen, N_2 = nitrogen, Cl_2 = chlorine, P_8 = phosphorus and S_8 = sulfur.

Chemical formula of compounds

A **compound** contains atoms of different elements. The chemical formula of a compound is a combination of the atomic symbols of the elements and numbers of atoms of each element that are present in the compound.

This lesson will only describe compounds containing atoms of two different elements. These are called binary compounds.

In some binary compounds, the atoms of different elements are present in equal numbers, for example sodium chloride (NaCl), magnesium oxide (MgO) and hydrogen chloride (HCl).

In other binary compounds the numbers of atoms of the two elements is not the same. Again, where there is more than one atom of an element, we show this using subscript numbers after their atomic symbols, for example carbon dioxide (CO_2), water (H_2O) and ammonia (NH_3).

Representing atoms and molecules

Atoms of different elements can be shown as spheres of different colours. The atoms are not

> **Fun fact**
>
> Using different colours to represent atoms only works for a small number of common elements. If you wanted to present atoms of all elements in this way you would need 98 different colours and it would become very complicated.

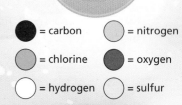

● = carbon   ○ = nitrogen
○ = chlorine   ● = oxygen
○ = hydrogen   ○ = sulfur

FIG 8.1.1 Representing atoms of some different elements

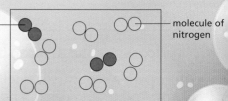

molecule of oxygen — molecule of nitrogen

FIG 8.1.2 Oxygen molecules and nitrogen molecules

really different colours but it helps to show which atoms are present in a molecule. Molecules of elements contain only one type of atom.

Molecules of compounds contain atoms of more than one element. Each molecule of hydrogen chloride contains one atom of hydrogen and one atom of chlorine (Fig 8.1.4).

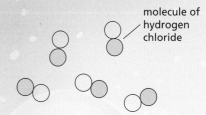

FIG 8.1.3 Sulfur molecule

Activity 8.1.1

Making molecules using balls and sticks

Here is what you need:

- Items that represent atoms, such as polystyrene balls or seeds

- Items that represent bonds between atoms, such as tooth picks

- Paints.

1. Decide how many atoms you need to make a model of an oxygen molecule.

2. Paint the atoms an appropriate colour. Join the atoms together to make a molecule of oxygen.

3. Decide how many atoms you need to make a model of an ammonia molecule.

4. Paint the atoms appropriate colours. Join the atoms together to make a molecule of ammonia.

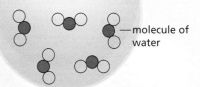

FIG 8.1.4 Hydrogen chloride molecules

FIG 8.1.5 Water molecules

Check your understanding

1. Copy and complete Table 8.1.1

Formula of the compound	Elements present in the compound	Number of atoms of each element
H₂O	hydrogen, oxygen	2 hydrogen atoms and 1 oxygen atom
HCl		
	carbon, oxygen	1 carbon atom and 2 oxygen atoms
SO₂		
	nitrogen, hydrogen	1 nitrogen atom and 3 hydrogen atoms
C₃H₈		

TABLE 8.1.1

Key terms

molecule two or more atoms chemically bonded together

compound substance composed of atoms of two or more substances

Combining power

We are learning how to:

- predict the formulae of compounds from the combining powers of their elements.

Combining power ⟩⟩⟩

The chemical formula of a compound shows the elements present and the relative proportions of each element in the compound.

The simplest compounds are formed from atoms of only two different elements, for example sodium chloride (NaCl), magnesium oxide (MgO), iron(II) sulfide (FeS) and carbon dioxide (CO_2). Other compounds contain atoms of more than two elements, for example sodium hydroxide (NaOH), calcium carbonate ($CaCO_3$) and copper sulfate ($CuSO_4$).

In order to determine the formula of a **binary compound** we need to know how many bonds the atoms of one element will form with the atoms of another. This is called the 'combining power' of the element. Values for common elements are given in Table 8.2.1.

Combining power		
1	**2**	**3**
Hydrogen (H)	Magnesium (Mg)	Aluminium (Al)
Lithium (Li)	Calcium (Ca)	Iron (Fe)
Sodium (Na)	Barium (Ba)	Nitrogen (N)
Potassium (K)	Copper (Cu)	Phosphorus (P)
Copper (Cu)	Iron (Fe)	
Chlorine (Cl)	Zinc (Zn)	
Bromine (Br)	Oxygen (O)	
Iodine (I)	Sulfur (S)	

TABLE 8.2.1 Combining of common elements

When using the information in Table 8.2.1 you should bear in mind that:

1. It is not possible to combine any two elements to make a compound. In general, compounds are formed between a metal and a non-metal, or two non-metals.

2. Where a compound contains a metal, this symbol is always given first.

3. In compounds of metals and non-metals, the ending of the non-metal changes to 'ide', for example oxide, chloride.

4. Some metals exhibit more than one combining power. To show this we give the combining power as a Roman numeral after the name of the metal, for example copper(I) and copper(II), iron(II) and iron(III).

A simple way to find the chemical formula of a binary compound is to write the combining power beneath each element and draw an appropriate number of arrows from one element to the other element.

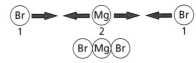

FIG 8.2.1 The chemical formula of potassium chloride is KCl

Activity 8.2.1

Finding the formulae of binary compounds

Here is what you need:

- 26 small pieces of card (4 cm by 4 cm).

1. Make cards to show the combining power of the elements in Table 8.2.1:

   a) Make one card for each metal with a combining power of 1.

   b) Make three cards for each non-metal with a combining power of 1.

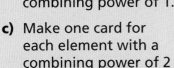

   FIG 8.2.3

   c) Make one card for each element with a combining power of 2 and of 3.

2. Use your cards to form as many compounds as you can, for example as shown in Fig 8.2.4.

3. Write the formula of each compound you make, for example AlCl₃.

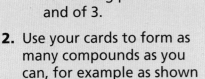

FIG 8.2.4

FIG 8.2.2 The chemical formula of magnesium bromide is MgBr$_2$

Check your understanding

1. Use the information in Table 8.2.1 to find the chemical formula of each of the following binary compounds.

   a) Hydrogen bromide
   b) Magnesium sulfide
   c) Aluminium iodide
   d) Iron(III) oxide

2. Each of the following formulae is incorrect. Use the cards from Activity 8.2.1 to work out the correct formula.

   a) KBr$_2$
   b) MgCl
   c) LiS
   d) AlO

Key terms

binary compound formed from the atoms of only two elements

combining power bonds an atom can make with other atoms

Ions

We are learning how to:

• describe the formation of ions from atoms.

 Ions 》》》

In the previous unit you may have wondered why different elements have different combining powers. The answer lies with the arrangement of electrons in their atoms.

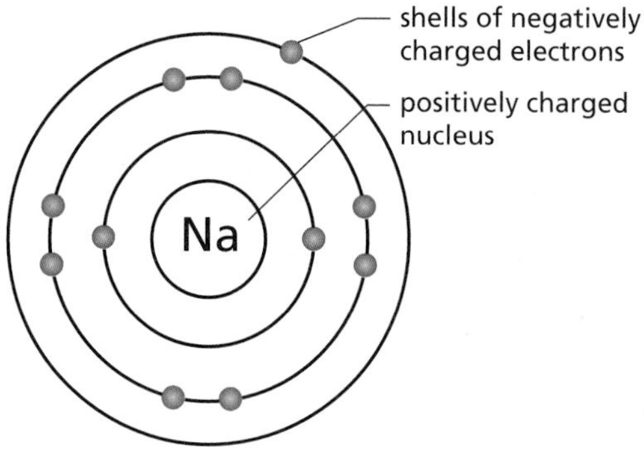

An atom consists of a positively charged nucleus surrounded by shells of negatively charged electrons. The electrons surrounding the nucleus of an atom can be classified into two groups according to their positions:

• The electrons in all but the outermost shell are called core electrons or inner electrons. They play no part in chemical reactions.

• The electrons in the outermost shell are called valence electrons. It is these electrons which are responsible for the combing power and chemical properties of an element.

Atoms become more stable when they have a full outermost shell. They can achieve this in two ways.

Elements whose atoms have only 1, 2 or sometimes 3 electrons in the outermost shell can lose these electrons. Since an electron carries a negative charge, losing electrons forms positively charged particles called **cations**.

Elements whose atoms only need to gain 1, 2 or sometimes 3 electrons to fill their outermost shell can gain electrons. Since an electron carries a negative charge, gaining electrons forms negatively charged particles called **anions**.

Compare Table 8.3.1 with Table 8.2.1 in the previous lesson.

Elements that form positively charged cations, which are mostly metals, combine with elements that form negatively charged anions, which are non-metals. There has to be sufficient of each type of **ion** to form a neutral compound.

shells of negatively charged electrons

positively charged nucleus

Na

FIG 8.3.1 Structure of a sodium atom

Elements that form positively charged ions or cations	Cation	Elements that form negatively charged ions or anions	Anion
Hydrogen	H^+	Chlorine	Cl^-
Lithium	Li^+	Bromine	Br^-
Sodium	Na^+	Iodine	I^-
Potassium	K^+		
Copper	Cu^+		
Magnesium	Mg^{2+}	Oxygen	O^{2-}
Calcium	Ca^{2+}	Sulfur	S^{2-}
Barium	Ba^{2+}		
Copper	Cu^{2+}		
Zinc	Zn^{2+}		
Iron	Fe^{2+}		
Aluminium	Al^{3+}	Nitrogen	N^{3-}
Iron	Fe^{3+}	Phosphorus	P^{3-}

TABLE 8.3.1 Ions formed by common elements

When atoms of elements have equal but oppositely charged ions, they combine in the ratio 1 : 1.

$Na^+ + Cl^-$ have equal but opposite charges, therefore the formula of sodium chloride is $NaCl$.

When atoms of elements carry different charges they combine in a ratio that forms a neutral compound.

Mg^{2+} and Br^- have opposite charges but the charge on the magnesium ion is twice that on the bromine ion. They combine in the ratio of 1 magnesium ion : 2 bromide ions, therefore the formula of magnesium bromide is $MgBr_2$.

K^+ and S^{2-} have opposite charges but to form a neutral compound they must combine in the ratio of 2 potassium ions : 1 sulfide ion. The formula of potassium sulfide is therefore K_2S.

Fun fact

Polyatomic ions contain atoms of more than two elements. Here are some common polyatomic ions that you may be familiar with.

Name	Formula
Ammonium	NH_4^+
Carbonate	CO_3^{2-}
Nitrate	NO_3^-
Sulfate	SO_4^{2-}

Activity 8.3.1

Making compounds by combining ions

Here is what you need:

- 26 small pieces of card (4 cm by 4 cm).

Here is what you should do:

1. Make cards to show the ions in Table 8.3.1. Make:

   a) one card for each positive ion

   b) three cards for each negative ion with a charge of −1

   c) one card for each of the other negative ions.

2. Use your cards to form as many compounds as you can, for example as in Fig 8.3.2.

FIG 8.3.2

3. Write the formula of each compound you make and the ions it contains, for example $MgBr_2$, Mg^{2+} and $2Br^-$.

Check your understanding

1. Use Table 8.3.1 to predict the formula of each of the following compounds.

   a) Hydrogen iodide   b) Calcium bromide

   c) Sodium sulfide   d) Aluminium oxide

Key terms

ion charged atom formed by the loss or gain of electrons

cation positively charged ion

anion negatively charged ion

Word equations and symbol equations

We are learning how to:

- write word equations and symbol equations to represent chemical reactions.

Equations >>>

A **chemical equation** describes what happens during a chemical reaction. The general form of a chemical equation is:

$$reactant(s) \rightarrow product(s)$$

The '**reactants**' are the chemical or chemicals we start with and the '**products**' are the chemical or chemicals we produce. Notice that an arrow '$\rightarrow$' is used in a chemical equation rather than the equals sign '=' used in a mathematical equation.

The simplest way to describe a chemical reaction is by a word equation. This gives the names of the reactants and products. The following equation represents a **synthesis reaction**:

$$iron + sulfur \rightarrow iron\ sulfide$$

Word equations are useful but they have their drawbacks.

In a symbol equation the names of the reactants and products are replaced by symbols:

$$Fe + S \rightarrow FeS$$

Symbol equations:

- are short to write
- can be understood by everybody because the same atomic symbols and formulae are used by scientists of all nationalities
- show the numbers of atoms and molecules involved in a reaction.

When learning to write symbol equations you should:

1. Write out a word equation for the reaction.

2. Replace the name of each reactant and product by symbols.

For example, when charcoal, which is essentially carbon, burns in air it combines with oxygen to form the gas carbon dioxide.

FIG 8.4.1 Carbon reacts with oxygen to form carbon dioxide

This is an example of a **combustion reaction**. It is also an **oxidation reaction** since carbon combines with oxygen. A large amount of heat is also released during this chemical reaction.

The word equation for this reaction is:

carbon + oxygen → carbon dioxide

The symbols for the reactants and product are: carbon = C, oxygen = O_2 and carbon dioxide = CO_2. The symbol equation for this reaction is therefore:

$$C + O_2 \rightarrow CO_2$$

FIG 8.4.2

Activity 8.4.1

Writing the chemical equation for a decomposition reaction

A **decomposition reaction** is one where one compound is broken down, often by the use of heat, into two or more substances.

Here is what you need:

- Copper(II) carbonate
- Tin lid
- Tripod
- Heat source
- Eye protection.

1. Place a small amount of copper(II) carbonate on a tin lid and stand this on a tripod. See Fig 8.4.2

2. Gently heat the copper(II) carbonate until no further change takes place.

3. What evidence is there that a chemical reaction has taken place?

4. Carbon dioxide gas has been given off and the solid that remains is copper(II) oxide. Write a word equation for this reaction.

5. The chemical formula of copper(II) carbonate is $CuCO_3$ and the chemical formula for copper(II) oxide is CuO. Write a symbol equation for this reaction.

Check your understanding

1. Write a word equation, and then a symbol equation, for each of the chemical reactions described below.

   a) Calcium carbonate decomposes on heating to form calcium oxide and carbon dioxide gas.

   b) Sulfur reacts with oxygen to form sulfur dioxide gas.

Key terms

chemical reaction process in which atoms of different elements rearrange themselves to form a new substance(s)

reactants starting materials in a chemical reaction

products produced by a chemical reaction

synthesis reaction two or more substances combine to form a single product

combustion reaction substance combines with oxygen, and heat is given out

oxidation reaction substance combines with oxygen

decomposition reaction one substance breaks down to form two or more products

Properties of ionic compounds

We are learning how to:

- describe the typical properties of ionic compounds
- identify ionic compounds from their properties.

Although the properties of all **ionic compounds** are not identical, they exhibit certain general features.

Solubility

Many, but not all, ionic compounds readily dissolve in water. You will learn more about solubility later in this unit.

Solubility in water is a good indicator that a substance may be an ionic compound, but it does not provide absolute proof.

FIG 8.5.1 Not all ionic compounds are soluble in water

Melting point

Ionic compounds generally have very high **melting points**. Table 8.5.1 gives the melting points of some common ionic compounds.

Ionic compound	Melting point / °C
Calcium sulfate	1460
Magnesium oxide	2852
Potassium bromide	734
Sodium chloride	801

TABLE 8.5.1

The melting point of magnesium oxide is higher than many metals. It is used industrially to line the inside of furnace walls.

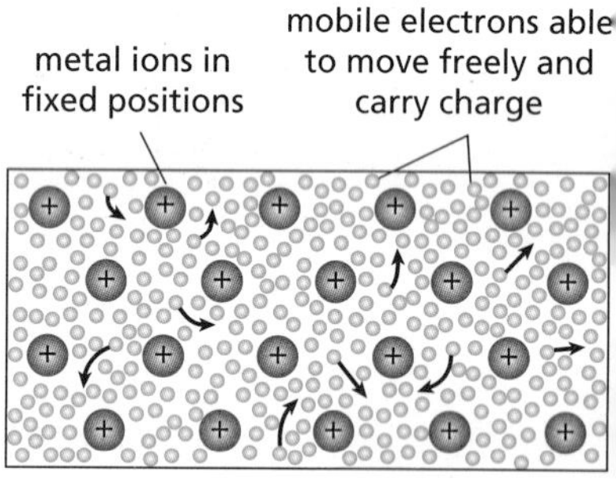

metal ions in fixed positions

mobile electrons able to move freely and carry charge

FIG 8.5.2 Electrons carry charge in a metal

Conducting electricity

An electric current is the result of charge being passed along a conductor.

In metals, charge is carried by **mobile** electrons. These are able to move through a **matrix** of positively charged metal ions.

Ions can also carry charge, but to do this they must be mobile.

Ionic compounds do not conduct electricity when solid even though they contain ions. They will only conduct electricity when the ions are mobile in aqueous solution or when the compound is molten.

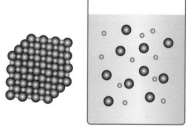

ions in fixed positions in solid

ions mobile when dissolved in water

○ sodium ion ● chloride ion

FIG 8.5.3 Ions carry charge when mobile

Activity 8.5.1

Identifying ionic compounds by conductivity

Your teacher will give you samples of different soluble compounds. Some of these will be ionic and some will not. You are going to investigate whether solutions of the compounds conduct electricity.

Here is what you will need:

- Battery of 3 cells
- Lamp
- Connecting wires
- Samples of different compounds
- Beaker 100 cm^3
- Stirring rod
- Distilled water.

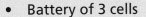

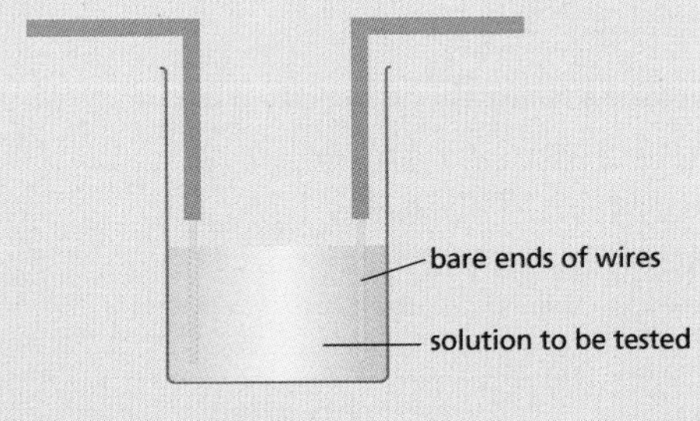

bare ends of wires

solution to be tested

FIG 8.5.4

> ### Fun fact
>
> Some ionic compounds are described as 'sparingly soluble'. This means that they are not insoluble but also that they are not very soluble. Calcium sulfate is sparingly soluble; the maximum amount that will dissolve in 1 dm^3 of water at room temperature is 2.4 g.

Here is what you should do:

1. Construct a circuit consisting of a battery, a lamp and a beaker into which each solution will be placed, all connected in series. Make sure the ends of the wires that go into the beaker have had the insulation removed.

2. Dissolve the first compound in a small amount of water in the beaker.

3. Place the ends of the wires into the beaker and see if the lamp lights up.

4. Wash the beaker and the ends of the wires.

5. Repeat steps 2–4 for all the compounds.

6. Which compounds were ionic and which were not?

Check your understanding

1. Table 8.5.2 contains information about compounds A and B.

Compound	Melting point/°C	Soluble in water	Conducts electricity as solid	Conducts electricity in solution
A	146	Yes	No	No
B	770	Yes	No	Yes

TABLE 8.5.2

State whether each of the compounds is ionic or not and give reasons.

Key terms

ionic compound substance composed of positive and negative ions

melting point temperature at which a solid becomes a liquid

mobile able to move position

matrix a framework

Electronic configuration

We are learning how to:

- write the electronic configurations of atoms of elements and the ions they form
- predict the electronic configurations of atoms of elements and the ions they form.

Electronic configuration of the first 20 elements 》》》

The arrangement of electrons in the atoms of an element is described as the **electronic configuration**. Table 8.6.1 shows the electronic configurations of the first 20 elements in the Periodic Table.

Element	Symbol	Number of electrons	Electron shells				Electronic configuration
			1st	2nd	3rd	4th	
Hydrogen	H	1	1				1
Helium	He	2	2				2
Lithium	Li	3	2	1			2,1
Beryllium	Be	4	2	2			2,2
Boron	B	5	2	3			2,3
Carbon	C	6	2	4			2,4
Nitrogen	N	7	2	5			2,5
Oxygen	O	8	2	6			2,6
Fluorine	F	9	2	7			2,7
Neon	Ne	10	2	8			2,8
Sodium	Na	11	2	8	1		2,8,1
Magnesium	Mg	12	2	8	2		2,8,2
Aluminium	Al	13	2	8	3		2,8,3
Silicon	Si	14	2	8	4		2,8,4
Phosphorus	P	15	2	8	5		2,8,5
Sulfur	S	16	2	8	6		2,8,6
Chlorine	Cl	17	2	8	7		2,8,7
Argon	Ar	18	2	8	8		2,8,8
Potassium	K	19	2	8	8	1	2,8,8,1
Calcium	Ca	20	2	8	8	2	2,8,8,2

TABLE 8.6.1 Electronic configuration of the first 20 elements

Electrons are added to the lowest energy shell until that shell is full, i.e. it contains the maximum number of electrons allowed. The maximum number of electrons allowed in the first shell is 2. This is called a **duplet**. The maximum number of electrons allowed in the second shell is 8. This is an **octet**.

An ion is formed when an atom either loses or gains one or more electrons. This doesn't alter the structure of the nucleus of the atom but it does alter the electronic configuration.

Na	→	Na⁺

$$Na \rightarrow Na^+$$

sodium atom   sodium ion
2,8,1   2,8

The atoms of elements which form positively charged ions lose electrons.

$$Cl \rightarrow Cl^-$$

chlorine atom   chloride ion
2,8,7   2,8,8

The atoms of elements which form negatively charged ions gain electrons.

> **Fun fact**
>
> When drawing a diagram of an atom, it is not necessary to show the structure of the nucleus. It is sufficient to state the numbers of protons and neutrons in the nucleus. However, the electron configuration should always be shown as a series of shells.

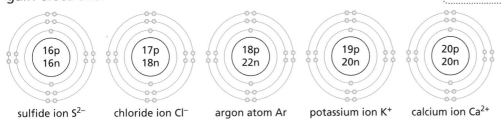

16p 16n	17p 18n	18p 22n	19p 20n	20p 20n
sulfide ion S²⁻	chloride ion Cl⁻	argon atom Ar	potassium ion K⁺	calcium ion Ca²⁺

FIG 8.6.1 All of these particles have the electronic configuration 2,8,8. They have 20 electrons.

Particles which have the same electronic configuration are described as **isoelectronic**.

Activity 8.6.1

Making models of isoelectronic species

The following are isoelectronic and have the electron configuration 2,8:

$$N^{3-} \quad O^{2-} \quad F^- \quad Ne \quad Na^+ \quad Mg^{2+} \quad Al^{3+}$$

Make models of these particles, showing the structure of their nuclei and their electronic configuration.

Check your understanding

1. Work out the electronic configurations of the following elements from the number of electrons given.

   a) Boron (5 electrons)   **b)** Chlorine (17 electrons)

   c) Nitrogen (7 electrons)   **d)** Argon (18 electrons)

2. Work out the electronic configurations of the following ions.

   a) Mg^{2+} **b)** F^- **c)** Al^{3+} **d)** S^{2-}

Key terms

electronic configuration arrangement of electrons around the nucleus of an atom or ion

duplet full electron shell containing 2 electrons

octet full electron shell containing 8 electrons

isoelectronic having the same electronic configuration

Law of conservation of mass

We are learning how to:

- explain that mass is neither created nor destroyed in a chemical reaction.

Law of conservation of mass ≫

Although there are many different possible chemical reactions, they all have one important feature in common.

During a chemical reaction no mass is lost or gained, therefore the total mass of the products is equal to the total mass of the reactants.

This is the **law of conservation of mass** and is true for all chemical reactions. It might sometimes appear that mass is lost during a chemical reaction but this is due to gaseous products escaping into the air.

Activity 8.7.1

Confirming the law of conservation of mass using a precipitation reaction

A **precipitation reaction** is one where two substances in solution combine to form products, one of which is a solid, and precipitates out of the solution.

Here is what you need:

- Potassium chloride solution
- Lead nitrate solution
- Beaker $100\,cm^3 \times 2$
- Measuring cylinder $25\,cm^3$
- Wash bottle containing distilled water
- Eye protection.

Here is what you should do:

1. Measure about $5\,cm^3$ of potassium chloride solution into a beaker using a measuring cylinder.

2. Thoroughly wash out the measuring cylinder with distilled water.

3. Measure about $5\,cm^3$ of lead nitrate solution into the second beaker.

4. Place the beakers onto a top-pan balance and record their total mass.

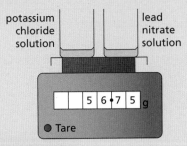

potassium chloride solution

lead nitrate solution

5 6•7 5 g

● Tare

FIG 8.7.1

5. Pour the contents of one beaker into the other and replace both beakers (one full and one now empty) on the balance. Record the mass of the products.

6. What evidence is there that a chemical reaction has taken place?

7. What evidence is there to support the law of conservation of mass?

8. The products of this reaction are lead chloride and potassium nitrate. Write a word equation for the reaction.

To confirm the law of conservation of mass for a chemical reaction in which one of the products was a gas, we would need to collect the gas and measure it.

For example, metal carbonates react with an acid to produce carbon dioxide gas. It isn't easy to physically measure the mass of a gas accurately; however, we can use the fact that 100 cm³ of carbon dioxide gas has a mass of 0.18 g at room temperature to work out the mass of a measured volume.

Check your understanding

1. The following word equation describes a chemical reaction in which a precipitate of silver bromide is formed:

silver nitrate + sodium bromide → silver bromide + sodium nitrate

a) If 1.70 g of silver nitrate was reacted with 1.03 g of sodium bromide, predict the mass of products formed.

b) If 0.85 g of sodium nitrate was produced, use the law of conservation of mass to predict the mass of silver bromide precipitated.

Key terms

mass amount of matter

precipitation reaction one in which an insoluble precipitate is formed

Balancing symbol equations

We are learning how to:

- balance symbol equations for chemical reactions.

Balancing equations ▶▶▶

When writing a symbol equation for a chemical reaction it is therefore important to ensure that there are equal numbers of atoms of each element on each side. This process is called balancing the equation:

$$\text{iron} + \text{sulfur} \rightarrow \text{iron sulfide}$$

$$Fe + S \rightarrow FeS$$

The above equation is already balanced since there is one atom of iron and one atom of sulfur on each side of the equation, but an additional step to balance an equation is often necessary.

Magnesium burns in air with a very bright flame to produce magnesium oxide:

$$\text{magnesium} + \text{oxygen} \rightarrow \text{magnesium oxide}$$

$$Mg + O_2 \rightarrow MgO$$

If you count the number of oxygen atoms on both sides of the equation you will see that an atom of oxygen has been lost. Perhaps we can correct this by writing a '2' on the right-hand side?

$$Mg + O_2 \rightarrow 2MgO$$

Now the number of oxygen atoms is the same but an atom of magnesium has been added! We must also write a '2' on the left-hand side:

$$2Mg + O_2 \rightarrow 2MgO$$

Now the equation is balanced. It has the same number of atoms of each element on both sides of the equations.

The following three steps will help you to write balanced chemical equations:

1. Write a word equation giving the names of the reactant(s) and the product(s).

2. Replace the names by atomic symbols and formulae.

3. If necessary, balance the equation by increasing the proportions of reactants and/or products.

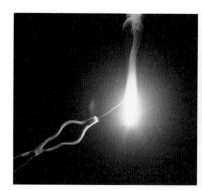

FIG 8.8.1 Magnesium burns brightly in air

FIG 8.8.2 Zinc reacts with dilute acids

Writing a balanced equation for the neutralisation reaction between sodium carbonate (Na_2CO_3) and dilute sulfuric acid (H_2SO_4)

A **neutralisation reaction** is one in which an acid and a base react to form neutral products.

Here is what you need:

- Sodium carbonate
- Dilute sulfuric acid
- Test tube
- Wooden splint
- Eye protection.

1. Place a small amount of sodium carbonate in a test tube.

2. Add enough dilute sulfuric acid to cover the sodium carbonate.

3. What evidence is there that a chemical reaction is taking place?

4. When there is no further reaction, light the end of a wooden splint and place it into the top of the test tube.

5. The gas produced during this reaction is carbon dioxide. State two properties of this gas based on your observations.

6. The word equation for this reaction is:

sodium carbonate + dilute sulfuric acid →
sodium sulfate + carbon dioxide + water

Write a balanced symbol equation for this reaction.

Check your understanding

1. Write chemical equations for the following reactions using the three-step method given above.

   a) Carbon (C) reacts with oxygen (O_2) to form carbon dioxide (CO_2).

   b) Copper (Cu) reacts with oxygen (O_2) to form copper(II) oxide (CuO).

   c) Magnesium (Mg) reacts with dilute hydrochloric acid (HCl) to form magnesium chloride ($MgCl_2$) and hydrogen (H_2).

   d) Zinc oxide (ZnO) reacts with dilute sulfuric acid (H_2SO_4) to form zinc sulfate ($ZnSO_4$) and water (H_2O).

   e) Sodium carbonate (Na_2CO_3) reacts with dilute nitric acid to form sodium nitrate ($NaNO_3$), carbon dioxide (CO_2) and water (H_2O).

Key terms
....

neutralisation reaction
an acid neutralises an alkali

Precipitation reactions and ionic equations

We are learning how to:
- identify precipitation reactions
- write ionic equations.

The **solubility** of a chemical compound is determined by how much of it will dissolve in a given volume of water (or some other solvent). Many chemical compounds are soluble in water while others are effectively insoluble. A general summary of the solubility of metal and ammonium compounds in water is given in Table 8.9.1.

From the table you can see that all compounds of ammonium, potassium and sodium are soluble in water, as are all nitrates.

Family of compounds	Soluble	Insoluble
Chlorides, bromides, iodides	All others	Lead and silver
Sulfates	All others	Barium calcium, lead
Carbonates	Ammonium, potassium, sodium	All others
Nitrates	All	

TABLE 8.9.1 Solubilities of different families of compounds

Precipitation reactions are carried out in solution. A **precipitate** is a solid that separates out from solution during a chemical reaction. In a typical precipitation reaction two soluble reactants form one soluble product and one insoluble product.

Precipitation reactions are useful because the products can be easily separated by filtration. When the reaction mixture is poured into a filter paper, the soluble product passes through as the **filtrate** while the insoluble product forms the **residue**.

You will see from Table 8.9.1 that silver iodide is an insoluble compound. It can be made by the reaction of a soluble silver compound with a soluble iodide. One possible combination is:

FIG 8.9.1 Formation of insoluble silver iodide

silver nitrate + potassium iodide → silver iodide + potassium nitrate

$AgNO_3$	KI	AgI	KNO_3
soluble	soluble	insoluble	soluble

Here is an ionic equation for this reaction:

$$Ag^+ + I^- \rightarrow AgI$$

Ionic equations make it easier to see exactly what species are involved in a chemical reaction.

- Spectator ions which do not take part in a reaction are removed.

- Ionic equations include only ions that take part in a chemical reaction.

Activity 8.9.1

Making silver bromide by a precipitation reaction

Here is what you need:

- Silver nitrate solution
- Potassium bromide solution
- Test tube × 2
- Boiling tube
- Filter funnel
- Filter paper
- Stand and clamp
- Eye protection.

1. Support a boiling tube using a stand and clamp.

2. Fold a filter paper and place it in a filter funnel. Place the filter funnel in the boiling tube.

3. Pour lead silver nitrate solution into a test tube until it is about one third full. Pour a similar amount of potassium bromide iodide solution into a second test tube.

4. Pour the lead silver nitrate solution into the test tube containing the potassium bromide solution.

5. What is the colour of the precipitate formed?

6. Pour the reaction mixture into the filter paper and leave it to separate.

7. Write a balanced symbol equation for this reaction.

8. Write an ionic equation for this reaction.

Check your understanding

Each of the following equations represents a precipitation reaction. The precipitate is written in **bold**. For each equation:

- Rewrite the equation showing the ions present in the reactants and the products.

- Delete the spectator ions.

- Write an ionic equation for the reaction.

1. $Ca(NO_3)_2 + Na_2CO_3 \rightarrow \mathbf{CaCO_3} + 2NaNO_3$

2. $Pb(NO_3)_2 + Na_2SO_8 \rightarrow \mathbf{PbSO_8} + 2NaNO_3$

Key terms

solubility amount of solid that will dissolve in a given amount of a solvent

precipitate insoluble solid formed during a chemical reaction

filtrate passes through a filter paper during filtration

residue retained by a filter paper during filtration

displacement reaction one element displaces another element from a solution of a compound

Exothermic and endothermic processes

We are learning how to:

- define the terms exothermic and endothermic
- identify examples of exothermic and endothermic processes.

Exothermic and endothermic reactions

The terms **exothermic** and **endothermic** are used to describe the loss or gain of energy during physical and chemical processes.

Processes in which energy is given out to the surroundings, such as the **combustion** of fuels, are described as exothermic. The energy is often in the form of heat.

Friction between the match head and the side of the matchbox provides a small amount of energy to light the match. Once the match is burning, it gives out lots of heat and light energy.

Processes in which energy is taken in from the surroundings are described as endothermic. **Photosynthesis** is an endothermic reaction in which energy, as sunlight, is absorbed.

If an endothermic reaction is carried out in a test tube, the test tube will feel cold because the reaction mixture has absorbed heat energy from the glass.

Changes of state

When ice is heated, it melts to become water. If the water is heated, it will boil to become steam. Since melting and boiling require energy from the surroundings, they are endothermic processes.

Conversely, when steam condenses to form water, and when water solidifies to form ice, energy is lost to the surroundings, so these are exothermic processes.

FIG 8.10.1 Combustion is an exothermic process

FIG 8.10.2 Friction provides heat energy

FIG 8.10.3 Photosynthesis is an endothermic process

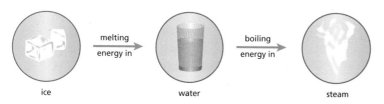

ice — melting energy in → water — boiling energy in → steam

FIG 8.10.4 Melting and boiling are endothermic processes

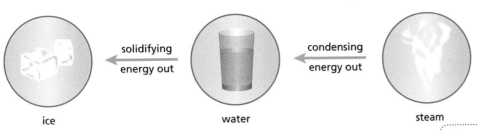

FIG 8.10.5 Condensing and solidifying are endothermic processes

Activity 8.10.1

Investigating exothermic and endothermic processes

Here is what you will need:

- Copper(II) sulfate solution
- Iron filings
- Ammonium chloride solid
- Distilled water
- Thermometer
- Boiling tube × 2
- Spatula.

Here is what you should do:

1. Pour copper(II) sulfate solution into a boiling tube to a depth of 2–3 cm. Place a thermometer into the copper(II) sulfate solution and record its initial temperature.

2. Add several spatulas of iron filings to the copper(II) sulfate. Stir and record the maximum temperature of the reaction mixture.

3. Pour distilled water into a boiling tube to a depth of 2–3 cm. Place a thermometer into the water and record its initial temperature.

4. Add several spatulas of solid ammonium chloride to the water. Stir and record the minimum temperature of the reaction mixture.

5. State whether each of the above was an exothermic or an endothermic process.

Check your understanding

1. In an experiment, a puddle of water was placed on a wooden board and a beaker was placed on top of it. A small amount of water was added to the beaker followed by solid ammonium nitrate. After stirring for a few minutes, it was seen that when the beaker was lifted the wooden base was stuck to it.

   Use your knowledge about what happens during an endothermic reaction to explain this observation.

FIG 8.10.7

- stirrer
- solid ammonium nitrate
- distilled water
- puddle of water
- wooden base

Fun fact

Chemical hand warmers contain chemicals that undergo an exothermic reaction. When the warmer is squeezed, the chemicals mix and heat is slowly given out.

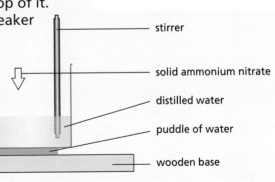

FIG 8.10.6

Key terms

exothermic giving out heat to the surroundings

endothermic taking in heat from the surroundings

combustion another word for burning

photosynthesis process by which green plants use energy from sunlight to make food

Review of Chemical bonding, formulae and equations

- A molecule consists of two or more atoms chemically bonded together.

- Some elements exist as molecules.

- A compound contains atoms of different elements.

- Binary compounds contain atoms of only two different elements.

- Compounds form between a metal and a non-metal, or two non-metals.

- Combining power is the number of bonds an atom of an element can make with other atoms.

- Ions are formed by the loss or gain of electrons from an atom.

- Positively charged ions are called cations.

- Negatively charged ions are called anions.

- Ions combine to form neutral compounds.

- Ionic compounds:
  o are sometimes but not always soluble in water
  o generally have high melting points
  o conduct electricity when in aqueous solution of molten.

- Electronic configuration is the arrangement of electrons around the nucleus of an atom.

- Electronic configuration changes when an atom forms an ion.

- A chemical equation describes what happens during a chemical reaction.

- A word equation identifies substances by name, and a symbol equation identifies them by their symbols and formulae.

- During a chemical reaction mass is neither created nor destroyed.

- In a balanced equation the same number of atoms of each element is present on the right-hand and left-hand sides of the equation.

- Equations cannot be balanced by altering the formulae of compounds.

- A precipitation reaction is one in which an insoluble solid product is formed.

- An ionic equation includes only the ions involved in a chemical reaction.

- Spectator ions are ions present during a chemical reaction but which take no part in it.

- During exothermic processes, heat is given out to the surroundings.

- During endothermic processes, heat is taken in from the surroundings.

Review questions on Chemical bonding, formulae and equations

1. Using the colour key given in Fig 8.1.1, draw each of the following molecules from the description given. State whether each is a molecule of an element or a compound.

   a) A molecule of hydrogen consists of two atoms joined together.

   b) A molecule of ammonia consists of three atoms of hydrogen joined to one atom of nitrogen.

   c) A molecule of chlorine consists of two atoms joined together.

   d) A molecule of tetrachloromethane consists of four atoms of chlorine joined to one atom of carbon.

   e) A molecule of hydrogen sulfide consists of two atoms of hydrogen joined to one atom of sulfur.

2. Write the formula of the ions formed by the following elements.

   a) Bromine   b) Magnesium   c) Sulfur

   d) Nitrogen   e) Lithium   f) Iron

3. Predict the formula of each of the following compounds.

   a) Potassium iodide   b) Water (hydrogen oxide)

   c) Magnesium nitride   d) Iron(III) oxide

4. Write a word equation and then a symbol equation for each of the reactions described below.

   a) Copper(II) carbonate decomposes on heating to form copper(II) oxide and carbon dioxide gas.

   b) Hydrogen reacts with chlorine to form hydrogen chloride gas.

   c) Barium reacts with iodine to form barium iodide.

   d) When iron(II) oxide is heated with aluminium, iron and aluminium oxide are formed.

5. Rewrite the following chemical equations so that they are balanced.

   a) $CuO + HCl \rightarrow CuCl_2 + H_2O$ b) $MgBr_2 + AgNO_3 \rightarrow AgBr + Mg(NO_3)_2$

   c) $Al + O_2 \rightarrow Al_2O_3$ d) $CH_4 + O_2 \rightarrow CO_2 + H_2O$

6. Describe the chemical reactions represented by each the following equations in words.

   a) $2Cu + O_2 \rightarrow 2CuO$ b) $Fe + 2HCl \rightarrow FeCl_2 + H_2$

   c) $ZnO + H_2SO_4 \rightarrow ZnSO_4 + H_2O$ d) $K_2CO_3 + 2HCl \rightarrow 2KCl + CO_2 + H_2O$

An aid for learning about bonding

Ionic compounds contain positively charged ions or cations, and negatively charged ions or anions. Cations and anions combine in small whole numbers to form compounds.

The salt that we sometimes sprinkle on the food we eat is an example of an ionic compound. Its chemical name is sodium chloride and it has the formula NaCl.

Sodium ions have the formula Na^+ and chloride ions have the formula Cl^- so they combine in the ratio of 1 : 1. Other cations, however, may have charges of 2^+ and 3^+ while other anions may have charges of 2^- and 3^-.

FIG 8.SIP.1 Salt is an example of an ionic compound

In order to make compounds (which are always neutral) it is sometimes necessary to combine cations and anions in ratios other than 1 : 1. This can be a bit confusing for students.

1. You are going to work in a group of 3 or 4 to produce a teaching aid that a teacher can use in the classroom to help students understand about bonding in ionic compounds. The tasks are:

    *   To look back at the unit content to make sure you understand about ions.
    *   To research how reference sources like books and the Internet explain the formation of ionic compounds.
    *   To plan your teaching aid and then make it.
    *   To try out your teaching aid on a group of students from another form and assess whether they find it helpful.
    *   To compile a presentation during which you will demonstrate how your teaching aid should be used.

    a) Look back through the unit and particularly lessons 8.2, which describes combining power, and 8.3, which describes ions. Make sure that you understand how cations and anions combine to form compounds.

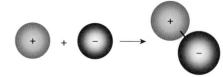

FIG 8.SIP.2 An example from a textbook of how coloured balls may be used to represent ions

    b) Look at how the formation of ionic compounds is illustrated in some textbooks. You have already carried out an activity in which different ions were represented by cards, and compounds were made by combining cards. Look at how the formation of ionic compounds is shown on the Internet. You might find that video sequences, models or cartoons are used.

    What are the good features of what you have seen that you would like to incorporate in your aid? For example, was colour used to good effect?

What are the bad features of what you have seen that you would like to avoid in your aid. For example, did things become too complicated too quickly?

c) How are you going to demonstrate how ions combine in a simple and entertaining way? Here is one example that might give you some ideas.

Here are some important features of the above you might consider:

- Cations and anions are represented by different colours.
- Cations have hooks and anions have eyes.
- The charge on the ion is represented by that number of hooks/eyes.

What features do you think are important in a good teaching aid? Here are some suggestions:

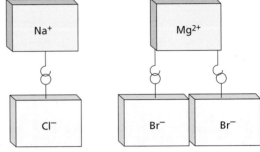

FIG 8.SIP.3 Representing ionic compounds using blocks

- Robust
- Easy to use
- Large enough to see clearly
- Material can be obtained easily or cheaply
- Materials are easy to work with.

Design your aid incorporating those features that you think are important.

Once you are happy with your design make a list of the materials and tools that you will need. Once you have collected what you need, make your aid.

d) Once your aid is built you need to try it out with some students from another form. How are you going to decide if it helped them understand ionic compounds or not?

Here are some ideas:

- You could interview them.
- You could ask them to complete a questionnaire. You will have to design this in advance of the trial. Use a word processor to do this.
- You could ask them to answer the same questions about ionic compounds before and after the trial. If your teaching aid has worked you might expect individual marks on the test to improve. You will need to write the questions in advance of the trial. Use a word processor to do this.

e) Prepare a PowerPoint presentation in which you describe your teaching aid and demonstrate how it should be used. You should discuss how you trialled your aid and how you assessed if it was successful or not. You should also describe how, as a result of the trial, your aid might be modified to make it even more effective.

Unit 9: Sensitivity and coordination

We are learning how to:

- describe the different parts of the nervous system.

The nervous system »

The nervous system is a network of nerve cells and fibres that carry nerve impulses between different parts of the body. The nervous system allows us to gather information about the environment immediately surrounding us, using our senses, and to act on it.

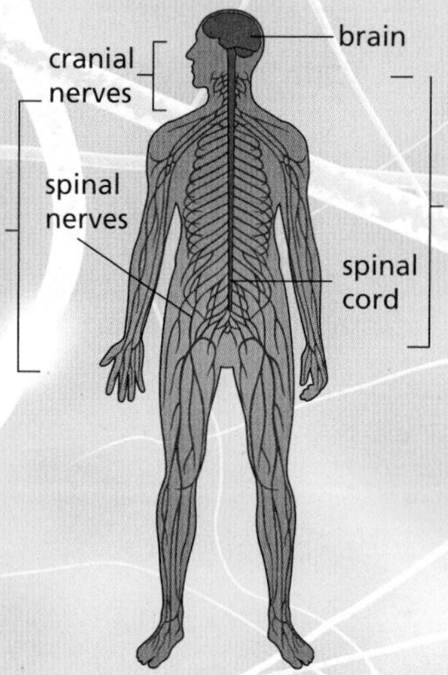

The nervous system is also essential in processes like movement and controlling body temperature.

There are two main parts to the nervous system:

- The **central nervous system (CNS)** is the brain and the spinal cord. This is the main control centre for the body.

- The **peripheral nervous system (PNS)** is a complex network of nerves that extends from the CNS to all parts of the body.

FIG 9.1.1 The nervous system

Any damage to the CNS will impair the ability of the body to function and might be fatal. The brain is protected by the skull while the spinal cord is contained within the vertebrae which make up the backbone.

People who have been in accidents where they have sustained damage to their spinal cord may lose the use of their arms or legs.

The PNS contains both sensory cells, called **receptors**, and **effectors**. The receptors are located in the sense organs such as the eyes and the ears. Their role is to gather information. This is often the result of a change in the environment called a **stimulus**.

FIG 9.1.2 Damage to the spinal cord can prevent the legs from working

The information is sent from the PNS to the CNS along sensory nerves. In the CNS decisions are made about how the body should respond. Messages are then sent from the CNS to the effectors in the PNS along motor nerves. Effectors are muscles and various glands which bring about the response.

How do you think your nervous system might react to the stimulus of a car coming around the bend as you are crossing a road?

FIG 9.1.3 Crossing a road

```
                    ┌─────────────────┐
                    │      CNS        │
                    │ brain and spinal cord │
                    └─────────────────┘
```

sensory nerves carry
information to the CNS

motor nerves carry
information from the CNS

PNS

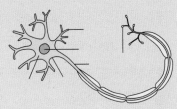

receptors	effectors
sense organs which detect stimuli	muscles and glands that bring about response

FIG 9.1.4 Information passes between the PNS and the CNS

Activity 9.1.1

Labelling a neuron

Nerve cells are called neurons. They look different from any other cells that you have already seen, like the red and white blood cells.

The following diagram shows a neuron. The labels point to the:

FIG 9.1.5 A neuron

axon, cell membrane, cytoplasm, dendrite,
nerve ending, nucleus

Here is what you should do:

1. Copy the diagram into your exercise book.

2. Add the labels by choosing from the words above. You might have to do some research to identify some of the parts.

Check your understanding

1. A student is out for a walk in the woods on a windy day when the following happens.

FIG 9.1.6

Describe how the student's nervous system might react to this situation.

Key terms

central nervous system brain and spinal cord

peripheral nervous system network of nerves around the body

receptor sensory cells that receive information

effector muscles and glands that respond to a stimulus

stimulus change in the environment

The sense organs

We are learning how to:

• describe the roles of the sense organs.

The sense organs

The sense organs are those parts of the body that respond to stimuli by sending a message to the brain. The sense organs allow us to gather information about our surroundings.

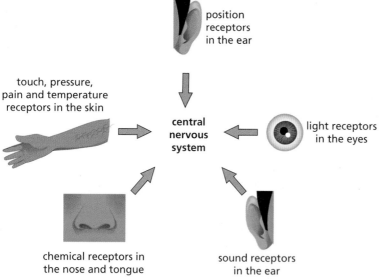

position receptors in the ear

touch, pressure, pain and temperature receptors in the skin

central nervous system

light receptors in the eyes

chemical receptors in the nose and tongue

sound receptors in the ear

FIG 9.2.1 The sense organs

FIG 9.2.2 Eyes detect light

FIG 9.2.3 Ears detect sound

Each sense organ responds to a different stimulus.

At the back of the eye there is a layer of light-sensitive cells called the retina. These cells detect light and send messages to the brain via the **optic nerve**.

The ear detects sound as vibrations that pass through the air. These vibrations are magnified by the ear and messages are carried to the brain by the **auditory nerve**.

FIG 9.2.4 Nose detects smell

Ears can detect very low-pitched sounds, like the rumbling of a truck, to very high-pitched sounds like the sound of a bird.

The inside of the nose is lined with cells that can detect different chemicals. We learn to associate different smells with the things that are familiar within our environment. For example, we might associate a fresh and slightly scented smell with flowers. Information is sent from the nose to the brain along the **olfactory nerve**.

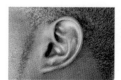

FIG 9.2.5 Tongue detects tastes

The tongue also detects chemicals in the foods that we eat and passes information to the brain. Different areas of the tongue detect different tastes. In some ways the skin is the most sophisticated of all the sense organs in that it can detect more than one type of stimulus. The skin contains receptors that detect being touched, whether an object is hot or cold, pressure and pain. All this information is passed on to the brain.

FIG 9.2.6 Skin detects a number of stimuli

Some areas of the skin, like the fingertips, have a much higher concentration of sense cells than other areas, like the skin on the back. This makes them more sensitive to stimuli.

Activity 9.2.1

Investigating touch receptors in the skin

You must work with a partner for this activity.

Here is what you need:

- Card
- Sticky tape
- Pins × 2.

Here is what you should do:

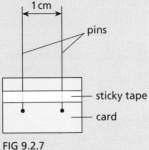
FIG 9.2.7

1. Attach the pins to a piece of card so that the points are about 1 cm apart as shown in Fig 9.2.7.

2. Your partner should not watch what you do so ask them to look away. Choose an area of their skin, such as their forearm.

3. Touch their skin (but don't stab them) with either one pin or two pins but don't tell them which.

4. Ask them whether they felt one pin or two pins and record whether they were correct or not.

5. Do this on the same area of skin four more times and record how many times out of five they were correct.

6. Repeat this by taking sets of five readings from different areas of skin. Write your results in the form of a table.

7. For which area(s) of the skin did your partner score the most correct answers?

8. For which area(s) of the skin did your partner score the least correct answers?

9. Which area contained the highest concentration of sensor cells and which area contained the lowest concentration of sensor cells?

Fun fact

The skin is the largest organ of your body. It is an important sense organ and contains receptors that can be stimulated by touch, hot and cold, pressure and pain.

Check your understanding

1. Copy and complete the following table.

Sense organ	What does the sense organ detect?
	Light
Ears	
Nose	
	Different tastes
Skin	

TABLE 9.2.1

Key terms

optic nerve carries nerve impulses from the eyes to the brain

auditory nerve carries nerve impulses from the ears to the brain

olfactory nerve carries nerve impulses from the nose to the brain

The human brain

We are learning how to:

- identify the different parts of the brain and their functions.

The human brain »»

The brain is the 'central processor' of the body. It gathers information from all of the sensory organs and then decides how the body should respond.

The brain can be divided into three main parts. Each of these parts has particular functions.

The **cerebrum** is the large part that you would normally see on a photograph of the brain. It is divided into two cerebral hemispheres and the surface is grooved.

The cerebrum is the part of the brain to which information is sent and stored. The brain uses this information to make decisions about voluntary actions. These are actions that we control; we can choose to do them or not.

The cerebrum also controls some functions such as speech, learning and memory.

The **cerebellum** is the region at the bottom of the brain. Its function is to coordinate muscle action so that we maintain our balance.

This allows us to stand and move about without falling or tripping ourselves up. This is described as involuntary action since it happens automatically; we have no control over it.

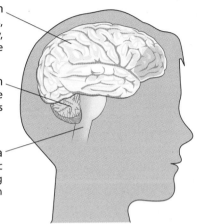

cerebrum
coordinates sensations, movements, memory, thought and intelligence

cerebellum
coordinates balance and precise movements

medulla oblongata
coordinates automatic processes, such as breathing and the circulation

FIG 9.3.1 The three main regions of the brain

FIG 9.3.2 The cerebrum is where we think

FIG 9.3.3 Good balance allows us to do things like dancing and sports

The **medulla oblongata** is also called the brain stem. It is the part of the brain that joins the spinal cord. This region controls a number of different functions including breathing, heartbeat, peristalsis (the movement of food along the alimentary canal) and digestion.

The medulla oblongata is also responsible for maintaining a constant body temperature. All of these actions are involuntary since we have no control over them.

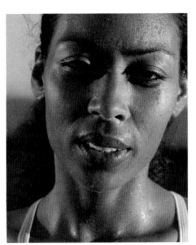

FIG 9.3.4 Temperature control is essential for our wellbeing

Activity 9.3.1

Identifying different parts of the brain

Here is what you need:

- An outline of the brain like the one shown in Fig 9.3.5.

Here is what you should do:

1. Label the skull and the vertebral column.

2. Lightly shade in and label the cerebrum, the cerebellum and the medulla oblongata.

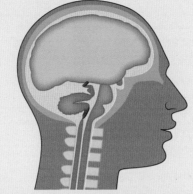

FIG 9.3.5 The three main regions of the brain

Check your understanding

1. In which parts of the brain do the following take place?

   a) Balance is coordinated.

   b) Information is stored.

   c) Breathing is controlled.

   d) Decisions are made.

   e) Body temperature is maintained.

Key terms

cerebrum largest part of the brain, where information is gathered and stored

cerebellum the part of the brain that coordinates muscle action

medulla oblongata the part of the brain that joins the spinal cord and which coordinates a number of essential functions

Voluntary, involuntary and reflex actions

We are learning how to:

- differentiate between voluntary actions, involuntary actions and reflex actions.

Voluntary, involuntary and reflex actions 〉〉

In the previous lesson you learnt that some parts of the brain are concerned with voluntary actions and other parts with involuntary actions.

- A **voluntary action** is one over which we have complete control. We can decide whether to do it or not.
- An **involuntary action** is one over which we have no control. The body takes this action without us having to think about it or make decisions. Many involuntary actions start when we are born and will continue to take place in the body until we die.

A **reflex action** is an involuntary response to a stimulus. Like all involuntary responses, it takes place without us thinking about it. Reflex actions help to protect the body in dangerous situations.

Accidentally touching a hot pan is painful and might burn the skin. Even in the short time it would take the messages to travel to and from the brain for a voluntary reaction, the skin on your hand would be damaged and it would feel very painful.

Instead of a voluntary action, your nervous system takes over and moves your hand without you having to think about it. This is called a reflex action. It takes place as a result of a **reflex arc**.

Here is how a reflex arc would prevent your fingers from being seriously damaged:

- The heat receptors in the hand send a message to the spinal cord.
- The reflex response is sent directly from the spinal cord to the muscles in the arm.

The brain is not involved in any decision making but a message is sent from the spinal cord to the brain so that you are aware of what your body is doing.

Blinking and the knee-jerk reflex are other examples of reflexes. Blinking keeps the surface of your eye moist and removes any particles of dust from it.

FIG 9.4.1 Running is a voluntary action

FIG 9.4.2 Breathing and heartbeat are involuntary actions

FIG 9.4.3 Touching a hot pan can be painful

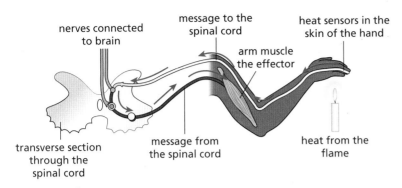

FIG 9.4.4 Reflex arc

The knee-jerk reflex is the body's response to tapping just under the knee cap to stretch the tendons.

Activity 9.4.1

Observing reflex actions

You must work with a partner for this activity. Both partners should observe the reflex actions in turn.

Here is what you should do.

1. Investigate the pupil reflex. The black centre of the eye is called the pupil. It is actually a hole through which light enters the eye.

2. Your partner should turn his or her back away from the window and cover their eyes with their hands for a few minutes. When they remove their hands look quickly at the size of their pupils. What do you observe?

3. Repeat this but with your partner looking towards a bright window. Once again, when they remove their hands look quickly at the size of their pupils. What do you observe?

4. Draw what you observed onto a pair of diagrams like the ones in Fig 9.4.5.

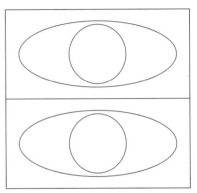

FIG 9.4.5

Check your understanding

1. This person is shivering.

   a) Give a stimulus that often causes people to shiver.

   b) Is shivering a reflex action? Explain your answer.

   c) How does shivering help the body?

FIG 9.4.6

Key terms

voluntary action an action that the brain controls

involuntary action an action that the brain doesn't control

reflex action involuntary response to a stimulus

reflex arc pathway taken by nerve impulses during a reflex action

Reaction time

We are learning how to:

- describe reaction time
- discuss factors that might affect reaction time.

Reaction time »»

The nervous system allows the body to respond to stimuli by taking voluntary actions. The time it takes to respond to a stimulus is called the reaction time.

Reaction time to a stimulus depends on:

- how quickly the brain can process the information provided by the stimulus and decide on a suitable response
- how quickly the muscles can act to bring about that response.

When we are young we are still learning about the world around us and our muscles are still developing. It is not surprising therefore that when children are young they have slow response times.

As children grow they become more knowledgeable and developed so they are able to respond more quickly to a stimulus.

As we grow we also build up a 'library' of stimulus–response situations that we experience regularly. People can reduce their reaction time through practice.

Cricketers practise batting in the nets (Fig 9.5.3). This allows them to respond more quickly to balls bowled at different speeds, heights and directions. They get used to playing different shots so they don't have to think about what to do.

Martial artists like Sheckema Cunningham must be able to react quickly to an attack by their opponent. They spend a lot of time practising different moves so they don't have to stop and think about them during bouts.

FIG 9.5.1 Young children have slow reaction times

FIG 9.5.2 Reaction times get shorter with age

FIG 9.5.3 Cricketers can respond very quickly

FIG 9.5.4 Martial artists practise moves

Fun fact

A cricket ball bowled by the fastest pace bowlers travels at up to 160 km/h. The batter has less than half a second to make up their mind what shot to play.

Measuring reaction time

You must work with a partner for this activity. Both partners should measure their reaction time in turn.

Here is what you need:

- 30 cm ruler.

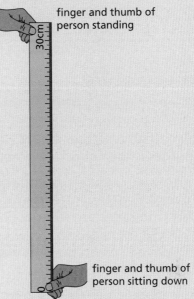

finger and thumb of person standing

30 cm

Here is what you should do:

1. One person should be seated and the other standing.

2. The person standing should hold the top of the ruler at the very end so that it is hanging in front of the person sitting.

3. The person sitting should place their thumb and finger level with 0 cm on the ruler, ready to catch it as it falls.

4. The person standing should release the ruler. As soon as he or she sees the ruler moving, the sitting person should bring their thumb and finger together to stop the ruler.

finger and thumb of person sitting down

FIG 9.5.5

5. The measurement on the ruler where the sitting person stopped it should be recorded to the nearest centimetre in a table like the one below.

6. This should be repeated five times and an average value taken.

	1st test	2nd test	3rd test	4th test	5th test
Value in cm					

TABLE 9.5.1

7. What was the average distance travelled by the ruler in the five tests?

8. What is the link between distance the ruler travels and reaction time?

Check your understanding

1. The graphs in Fig 9.5.6 show how average reaction time varies with age and gender.

   a) Does the diagram suggest that men or women have the quicker reaction time?

   b) Is it true that a man of 70 has a similar reaction time to a child of 5 years old?

   c) Suggest how a person might prevent their reaction time from increasing as they grow older.

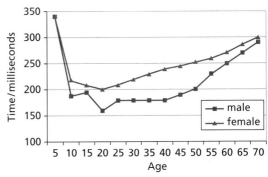

FIG 9.5.6

The endocrine system

We are learning how to:

- identify the glands in the endocrine system
- describe the roles of the glands in the endocrine system.

The endocrine system

The nervous system is one way in which messages can be sent from one part of the body to another. A second means of sending messages is the endocrine system.

The endocrine system works in a different way to the nervous system. Instead of nerve impulses, messages are sent around the body as special chemicals called **hormones**. These are sometimes described as chemical messengers.

Each endocrine gland produces one or more hormones. When a hormone is released into the bloodstream it acts on one particular organ called the **target organ**. Each hormone has a specific role in the body.

The **pituitary gland** is a pea-sized structure found at the bottom of the brain. It is sometimes called the 'master gland' because it controls the release of hormones by other endocrine glands. The pituitary gland secretes a number of different hormones, including ones that:

- regulate the growth of the body
- control sperm production in the male and egg production in the female
- control the amount of water in the body.

The **thyroid gland** is found in the throat and its action is controlled by the pituitary gland. It secretes a hormone that controls the rate of metabolism.

The **pancreas** is connected to the small intestine. It secretes hormones that control the concentration of glucose in the blood. This is sometimes described as the blood sugar level.

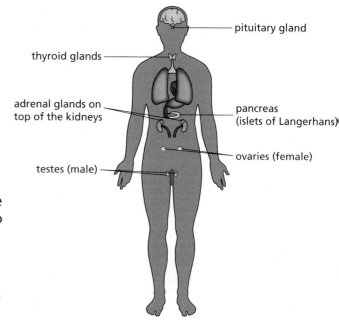

pituitary gland

thyroid glands

adrenal glands on top of the kidneys

pancreas (islets of Langerhans)

ovaries (female)

testes (male)

FIG 9.6.1 The endocrine system

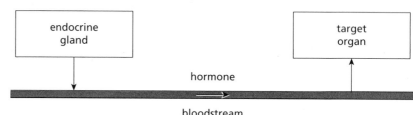

endocrine gland

target organ

hormone

bloodstream

FIG 9.6.2 Hormones are carried in the blood

The **adrenal glands** are located on the tops of the kidneys. They produce hormones that prepare the body for sudden actions or emergencies.

The male **testes** and the female **ovaries** both produce hormones that allow the body to develop secondary sexual characteristics as a person grows to become an adult.

The hormones released by the ovaries also control the menstrual cycle.

Activity 9.6.1

Marking the locations of the endocrine glands

Here is what you need:

- An outline of the human body like the one in Fig 9.6.3.

Here is what you should do:

1. Mark the position of each endocrine gland on the outline. You should mark the endocrine glands associated with the reproductive organs according to whether you are a boy or a girl.

2. Label each endocrine gland.

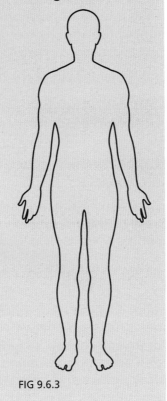

FIG 9.6.3

Check your understanding

1. **a)** In which part of the body is the thyroid gland?

   b) Which endocrine gland is sometimes called the master gland?

   c) Which endocrine glands produce different hormones in men and in women?

   d) Which endocrine gland produces hormones that control the concentration of glucose in the blood?

Fun fact

In the body some glands have ducts or tubes along which substances pass. The endocrine glands do not have ducts and are therefore described as ductless glands. Any substance they produce goes directly into the bloodstream.

Key terms

hormone chemical messenger released by an endocrine gland

target organ the organ upon which a hormone will act

pituitary gland endocrine gland at the base of the brain

thyroid gland endocrine gland in the throat

pancreas endocrine gland attached to the small intestine

adrenal glands endocrine glands on top of the kidneys

testes part of the male reproductive system that releases hormones

ovaries part of the female reproductive system that releases hormones

Hormones

We are learning how to:

- explain what hormones are
- describe the roles of some hormones in the human body.

Hormones

Each endocrine gland is responsible for secreting particular hormones.

The pituitary gland releases a number of different hormones. Some of these control the release of hormones by other endocrine glands. Hormones produced by the pituitary gland include:

- growth hormone, which regulates the growth of the body
- **antidiuretic hormone (ADH)**, which controls the reabsorption of water by the kidneys during the production of urine
- reproductive hormones, which control the production of sperm (in males) and ova (in females).

The thyroid gland releases **thyroxine**, which controls the rate of body metabolism. The release of this hormone is in turn controlled by the pituitary gland.

The pancreas releases **insulin** and **glucagon**, which control the level of glucose in the blood.

The adrenal gland releases **adrenalin**, which is sometimes called the 'fight or flight' hormone. It has a number of effects which prepare the body for an emergency, including: increasing the rate of heartbeat and breathing, causing the concentration of glucose in the blood to rise and increasing the supply of blood to the brain and muscles.

The testes release **testosterone**, which develops male secondary sexual characteristics like the development of facial hair and pubic hair, a deepening of the voice and an increase in body size.

The ovaries release:

- **oestrogen**, which develops the female secondary sexual characteristics like the growth of pubic hair, enlargement of breasts and broadening of the hips
- oestrogen and progesterone, which together regulate the menstrual cycle.

The action of a hormone depends on its concentration in the blood.

Fun fact

A hormone imbalance is a condition where a person's body produces too much or too little of a hormone. Both of these can have serious implications for the person. For example, the hormone thyroxine controls the rate of body metabolism. Too little of this and the metabolism slows down, producing tiredness. Too much and the person becomes thin as the body uses up all its energy reserves.

FIG 9.7.1 This man has had a surge of adrenalin

FIG 9.7.2 A hot day at Montego Bay

When the weather is hot the body loses water by sweating. The pituitary gland releases a high concentration of ADH into the blood so that the kidneys will retain more water. On a cold day the body loses less water by sweating. The pituitary gland releases a low concentration of ADH so that less water is retained by the kidneys.

Sometimes two hormones work together to control a process in the body. The concentration of glucose in the blood (blood sugar level) is controlled by the hormones insulin and glucagon.

- High levels of blood glucose cause the pancreas to secrete insulin, which converts glucose to a substance called glycogen for storage in the liver.

- Low levels of blood glucose cause the pancreas to secrete glucagon, which converts glycogen into glucose.

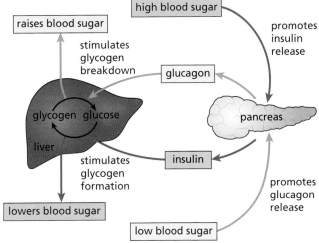

FIG 9.7.3 Controlling glucose concentration in the blood

Researching diabetes

This is a research activity. You should use whatever reference material, including the Internet, that you have available.

Diabetes is a condition in which the body needs help controlling the level of glucose in the blood. You might have a relative or friend who has this condition. Jamaica is one of 24 countries in the International Diabetes Federation (IDF) North America and Caribbean (NAC) region.

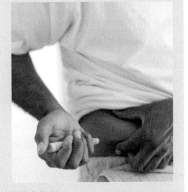

FIG 9.7.4

Find out what you can about diabetes in preparation for a class discussion.

Check your understanding

1. Name the hormone responsible for the following actions.

   a) Controlling the amount of water retained by the kidneys.

   b) Controlling the rate of metabolism in the body.

   c) Regulating a woman's menstrual cycle.

Key terms

antidiuretic hormone (ADH) hormone that controls water retention

thyroxine hormone that controls the rate of metabolism

insulin hormone involved in the control of blood sugar level

glucagon hormone involved in the control of blood sugar level

adrenalin hormone that prepares the body for emergencies, the 'fight or flight' hormone

testosterone hormone responsible for male secondary sexual characteristics

oestrogen hormone responsible for female secondary sexual characteristics

Comparing the nervous system and the endocrine system

We are learning how to:

- describe the differences between the nervous system and the endocrine system.

Comparing the nervous and endocrine systems　》》》

Although the **nervous system** and the **endocrine system** are both concerned with sending messages from one part of the body to another, there are certain important differences between the two systems.

Nerve messages are electrical impulses that travel along nerve cells or neurons rather like an electric current passes along the wires in an electric circuit. The nervous system extends throughout the whole body so the brain can pass a message to different muscles almost instantly.

Different types of nerve impulses travel at different speeds but they all travel very quickly. Impulses to the brain from touch sensors travel at around 75 m/s, while from pain sensors the speed is much slower, at around 1 m/s. The speed of impulses travelling in the opposite direction from the brain to muscles may be over 100 m/s.

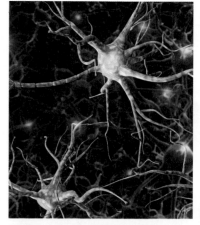

FIG 9.8.1 Nerve cells or neurons

Nerve impulses are ideal in situations where the body needs an immediate or short-term response. They are usually the response to a stimulus of one sort or another.

Blood vessels are hollow tubes along which blood flows. The blood moves around your body in a similar way to water flowing through the pipes in your home.

Blood moves around the body far more slowly than nerve impulses move along nerve cells. The speed of blood in arteries is around 50 cm/s while the speed of blood in veins, which don't experience the same pressure from the heart, is only 15 cm/s.

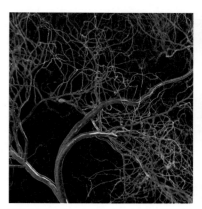

FIG 9.8.2 Blood vessels

Clearly, hormones are not able to provide an immediate or short-term response to a situation in the same way as nerve impulses. Hormones have a different role in the body. They bring about long-term changes.

Hormones are usually concerned with gradual change, and with keeping the internal environment of the body stable. For example, controlling the water content of the body or controlling the level of glucose in the blood. These are very important to our wellbeing but they don't require an instant change.

Activity 9.8.1

Is body temperature controlled by the nervous system or the endocrine system?

You will need to carry out some research for this activity. Use the Internet and/or whatever resources are available.

Maintaining a constant body temperature is essential to our wellbeing.

1. Is this something that you think involves short-term responses or is it more of a long-term change? What are your first thoughts? Do you think it is controlled by the nervous system or the endocrine system?

2. Use whatever reference material is available to find out whether you are correct or not.

> **Fun fact**
>
> The heart pumps about 7600 dm³ of blood each day, which is about 5 dm³ per minute. This is about the same as the amount of blood in your body. So if you were able to put a marker into your blood as it left your heart it would reappear coming back to your heart a minute later.

Check your understanding

1. Copy and complete Table 9.8.1, which compares the nervous system with the endocrine system.

Feature	Nervous system	Endocrine system
Nature of the message		
Way in which message is carried		
Speed of message		
Speed of response to message		
Reason for message		

TABLE 9.8.1

Key terms

nervous system brain, spinal cord and network of neurons along which nerve impulses are sent

endocrine system collection of glands that secrete hormones into the bloodstream

Review of Sensitivity and coordination

- The nervous system is a network of nerve cells that carry messages around the body.
- There are two main parts to the nervous system: the central nervous system (CNS) and the peripheral nervous system (PNS).
- The CNS is made up of the brain and spinal cord.
- The PNS is a network of nerve cells or neurons.
- Receptors are located in sense organs and respond to stimuli.
- Effectors are muscles which bring about a response.
- There are five sense organs: eyes, ears, tongue, nose and skin.
- Each sense organ responds to a different stimulus.
- The brain is the 'central processor' of the body and consists of three main parts.
- The cerebrum receives and stores information, and uses the information to make voluntary decisions.
- The cerebellum is concerned with balance.
- The medulla oblongata controls a number of body functions including breathing, heartbeat, peristalsis and digestion.
- A voluntary action is one over which we have complete control.
- An involuntary action is one over which we have no control.
- A reflex action is an involuntary response to a stimulus.
- Reflex reactions do not involve the brain.
- Reflex actions may prevent the body from injury.
- Reaction time is the time taken for the body to respond to a stimulus.
- Reaction time increases with age.
- Reaction time can be reduced by practice.
- The endocrine system consists of endocrine glands that secrete hormones into the bloodstream.
- Hormones are chemical messengers.
- A hormone secreted by an endocrine gland causes a response at a target organ.
- Endocrine glands include the pituitary gland, thyroid gland, pancreas, adrenal gland, testes and ovaries.
- Each endocrine gland secretes one or more hormones.
- Nerve impulses travel very quickly and bring about a short-term response.
- Hormones travel more slowly in the blood and generally bring about long-term changes.

Review questions on Sensitivity and coordination

1. What sense organ makes it possible to do the following?

   a) Hear music

   b) Sense that an object is cold

   c) See coloured lights

   d) Smell food cooking

   e) Taste different types of food.

2. **a)** Explain the difference between a voluntary and an involuntary action.

   b) In which part of the brain do we make decisions about voluntary actions?

   c) Give an example of an involuntary action that takes place in the cerebellum.

3. **a)** Describe an example of a reflex action.

   b) Why are reflex actions beneficial to the body?

   c) Why are reflex actions triggered by the spinal cord and not the brain?

4. The bar chart in Fig 9.RQ.1 shows how reaction time is affected under different circumstances when a person is driving a car.

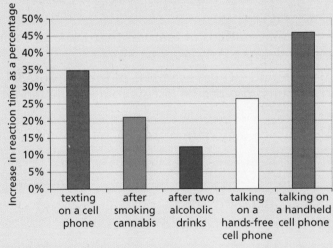

FIG 9.RQ.1

   a) Why is it important that a car driver has short reaction time in the event of an emergency?

   b) Which of the circumstances in the chart increases the reaction time most?

   c) Drinking two alcoholic drinks increases reaction time by the smallest percentage. Does this mean it is safe to drink alcohol and drive? Explain your answer.

   d) Predict the effect on reaction time of a person texting on a cell phone after having smoked cannabis.

Ensuring documents are legible

An important feature of our ability to detect light is that it allows us to read. Some people are born with defective eyesight. Most people find that their eyesight deteriorates with age.

To some extent, defective sight can be corrected using spectacles but choice of font size, font colour and background colour may also help.

The Government is concerned that people are not able to read the text on official documents effectively because of poor design. It is your job to provide advice on this.

FIG 9.SIP.1 The ability to read deteriorates with age

1. You are going to work in a group of 3 or 4 to investigate the clearest way to provide printed information. The tasks are:

   - To consider the different parameters associated with printed material. These include:
     - ○ choice of font
     - ○ font colour
     - ○ font size
     - ○ background colour.
   - To design some printed text that can be used for testing.
   - To devise methods of testing different texts and a means by which they can be graded in some way.
   - To make recommendations of certain combinations of features that work well and to identify others that should be avoided because they work badly.
   - To compile an oral presentation. This should include examples of the texts used, how they were tested and a summary of results. You should discuss your recommendations on the basis of your test results.

   a) There are many different fonts available on a modern word processor. Some of the fonts in common use are:

   Jamaica   Jamaica   Jamaica   Jamaica   Jamaica

   The words below are written in increasing font size. Is there a point beyond which increasing the font size has no effect on the readability of the text?

   Jamaica   Jamaica   Jamaica   Jamaica   Jamaica

   Traditionally we think in terms of black text on a white background but this 'norm' was established when there were few viable alternatives. With modern word processing it is very easy to write text in a coloured font on a contrasting coloured background.

   Are different coloured fonts equally easy to read?

   Jamaica   Jamaica   Jamaica   Jamaica   Jamaica

   Is the background colour important when reading text?

   Jamaica   Jamaica   Jamaica   Jamaica   Jamaica

   What is the effect of different coloured fonts on different coloured backgrounds?

b) You do not have time to test every possible combination of fonts, sizes and colours.

Your first job will be to identify a small number of each. You might play about on a word processor with different fonts, sizes and colours until you hit on combinations that you think work really well.

You might decide to focus on 4 fonts, 4 font sizes, 4 font colours and 4 background colours.

How are you going to test different combinations of font, size, colour and background effectively? One way might be to prepare short extracts of text presented in different ways. The extracts should be long enough for you to time how long a person takes to read it, so perhaps enough for 1 or 2 minutes.

Perhaps the extracts should all be different so the person reading them doesn't get used to the content and skip words. If the extracts are different, how are you going to ensure they are equal in demand with regard to language level? Text with lots of long words will be more difficult to read irrespective of font size and colour. You could choose a novel and take different lines of text from it.

c) How are you going to carry out your tests? You could aim to obtain both quantitative and qualitative data.

- You could obtain quantitate data by timing how long it takes a person to read an extract.

- You could obtain qualitative data by asking the person how easy or difficult they found the reading. They could score a mark between 0 for the most difficult and 5 for the easiest.

Who are you going to carry out your tests on? If there are some old people living nearby, you might ask them to be your guinea pigs. Failing that, you might ask students from another form if they will help.

d) How are you going to record and present the data you collect? You could do this on a spreadsheet. Are you going to test the same text with different readers and take average scores?

e) How are you going to decide how well or how badly each extract performs? If an extract is easy to read it should take a short time, while text which is more difficult will take longer.

f) Prepare an oral presentation in which you explain why you chose to explore particular combinations of font, size and colour. You should show some samples of the text you prepared and invite your audience to suggest which they find easiest and hardest to read. It will be interesting to compare the views of your audience with the results you obtained.

Describe the tests and detail the results you obtained. Explain how you are using these results to make your recommendations to the Government.

Unit 10: Acids and alkalis

We are learning how to:

- identify acids and alkalis as two groups of chemicals.

Acids and alkalis　》》

One way of classifying matter is into acids and alkalis. Acids and alkalis are found in a wide variety of products.

FIG 10.1.1 a) Fruit contains acid

b) Baking soda is an alkali

FIG 10.1.2 These household chemicals may be acids or alkalis and many of them are corrosive

Many acids and alkalis may be hazardous.

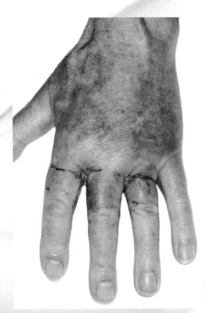

FIG 10.1.3 The effects of contact with sulfuric acid

The picture shows some of the effects of contact with sulfuric acid.

Calcium hydroxide is a mildly corrosive alkali. However, the product of the reaction of calcium hydroxide and sulfuric acid is calcium sulfate, which is a useful substance – it is used to make casts for supporting fractured bones.

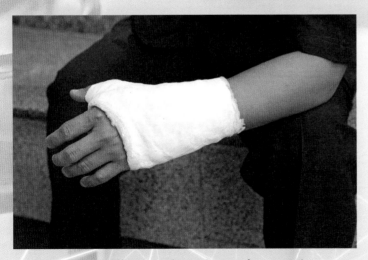

FIG 10.1.4 Calcium sulfate is used to make casts for supporting fractured bones

In this section you will be exploring how to distinguish between acids and alkalis. You will also learn about some reactions of acids.

Identifying acids

We are learning how to:

- distinguish between substances that are acids and alkalis
- explain the use of an indicator
- give the colour change in an indicator when it contacts an acid.

Identifying acids ⟫⟫

Activity 10.2.1

Exploring how indicators change colour with different acids

Here is what you need:

- Hydrochloric acid
- Sulfuric acid
- Nitric acid
- Red and blue litmus paper
- Phenolphthalein
- Methyl orange
- Test tubes
- Tweezers
- Droppers.

⚠️ **SAFETY**

Observe the safety icon on the acid bottles. All indicator papers should be held with tweezers. Avoid spillage.

Here is what you should do:

1. Copy Table 10.2.1.

Acid	Indicator	Colour change
hydrochloric (HCl) sulfuric (H_2SO_4) nitric (HNO_3)	red litmus paper	
hydrochloric sulfuric nitric	blue litmus paper	
hydrochloric sulfuric nitric	phenolphthalein	
hydrochloric sulfuric nitric	methyl orange	

TABLE 10.2.1

2. Pour a few drops of hydrochloric acid into each of two test tubes.

3. Suck up a dropperful of hydrochloric acid from one of the test tubes.

4. Squeeze one drop of hydrochloric acid onto a piece of each of the coloured indicator papers and record the colour you observe.

5. Pour one drop of the liquid indicators into each of the two test tubes of hydrochloric acid. Record the colour change you observe.

6. Wash out all the test tubes and droppers thoroughly.

7. Repeat steps 2 to 6 for sulfuric acid and then for nitric acid.

8. Compare the results you found for the various acids. Did all acids give the same results for each indicator?

The word **acid** is from the Latin *acidus*, meaning sour. An acid is a chemical substance. Solutions formed from these chemical substances usually have a sour taste. In order to identify acids, **indicators** are used. Indicators are made from special dyes and the results of your experiments should show the colour change.

> ### Fun fact
>
> Most fruits contain acids, such as citric acid, but these are weak acids and we can eat them without coming to any harm. They also provide the tangy flavours of fruits.

Activity 10.2.2

Check your breath

In this activity you will explore the gas that is exhaled from your body.

Here is what you need:

- Two test tubes
- Dropper
- Universal indicator
- Straw.

Here is what you should do:

1. Add a little water to two clean test tubes.

2. Put one drop of universal indicator into one of the test tubes and observe its colour.

3. Place a straw into each test tube.

4. Ask a volunteer to gently breathe out a few breaths through the straw and into each test tube.

5. Is there a colour change in the indicator? What conclusions can you draw from this?

FIG 10.2.1

Carbon dioxide is a naturally occurring gas that is important for photosynthesis. It is produced during respiration, decay, fermentation, combustion and volcanic eruptions. Carbon dioxide is acidic and forms carbonic acid with water.

Check your understanding

1. What colour changes are observed when the following are added to an acid?

    a) Blue litmus paper
    b) Phenolphthalein
    c) Methyl orange.

2. Name three common acids.

Key terms

acid a type of chemical substance

indicators substances made from special dyes that change colour depending on whether an acid or alkali is present

Strength of an acid

We are learning how to:

- distinguish between substances that are acids and alkalis
- identify the strength of an acid.

Strength of an acid 》》

Activity 10.3.1

Exploring the strength of acids

Here is what you need:

- Vinegar
- Milk
- Apple
- Universal indicator solution

- Small pieces of universal indicator paper
- Tweezers
- Spatulas
- Scalpels

- Experiment trays
- Hydrochloric acid
- Sulfuric acid
- Nitric acid
- Droppers.

 SAFETY

Observe the safety icon on the acid bottles. All indicator papers should be held with the tweezers. Take care when using scalpels. Avoid spillage.

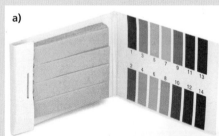

Here is what you should do:

1. Copy Table 10.3.1. List the name of the substances you are going to test in the Specimen column, as shown below.

FIG 10.3.1 **a)** Universal indicator paper **b)** Universal indicator solution **c)** Experiment trays

| Specimen | Universal indicator (colour change) | |
	Paper	Solution
vinegar		
milk		
apple		
hydrochloric acid		
sulfuric acid		
nitric acid		

TABLE 10.3.1

2. Using a scalpel, spatula or dropper, place a tiny sample of each specimen in the cavities of the experiment tray.

3. Wash your hands and all the apparatus you used thoroughly, and dry your hands.

4. Place one little piece of indicator paper in each specimen. Observe and record the colour change.

5. Add one drop of indicator to each specimen. Observe and record the colour change.

6. Were all the colours the same?

7. Do you think that the difference in colour has any significance for acid strength?

Universal indicator is very special. As well as indicating acidity, it also gives the strength of an acid.

FIG 10.3.2 As acidity weakens, the colour of universal indicator moves from red to yellow

Fun fact

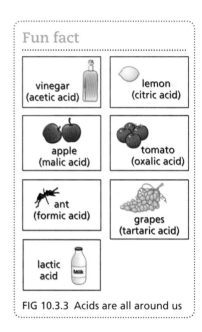

FIG 10.3.3 Acids are all around us

The colours of universal indicator match those on a range called the pH scale. Acids can have a **pH** with a number between 0 and 6. The stronger the acid, the more corrosive it is. The stomach produces hydrochloric acid, which is the strongest of all acids.

Check your understanding

1. How can you determine the strength of an acid?

2. Research what scientists use to make indicators.

3. Research what can be used to make homemade indicators.

Key term

pH measure of strength of an acid or alkali

Identifying alkalis

We are learning how to:

- distinguish between substances that are acids and alkalis
- give the colour change in an indicator when it is in contact with an alkali.

Identifying alkalis 》》

Activity 10.4.1

Exploring alkalis

Here is what you need:

- Sodium hydroxide
- Aqueous ammonia
- Calcium hydroxide
- Blue litmus paper
- Red litmus paper
- Phenolphthalein
- Methyl orange
- Test tubes
- Tweezers
- Droppers.

 SAFETY

Observe the safety icon on the alkali bottles. All indicator papers should be held with the tweezers. Avoid spillage.

Here is what you should do:

1. Copy Table 10.4.1.

Alkali	Indicator	Colour change
sodium hydroxide (NaOH) aqueous ammonia (NH$_4$OH) calcium hydroxide (Ca(OH)$_2$)	blue litmus paper	
sodium hydroxide aqueous ammonia calcium hydroxide	red litmus paper	
sodium hydroxide aqueous ammonia calcium hydroxide	phenolphthalein	
sodium hydroxide aqueous ammonia calcium hydroxide	methyl orange	

TABLE 10.4.1

2. Pour a few drops of sodium hydroxide into each of two test tubes.

3. Suck up a dropperful of sodium hydroxide from one of the test tubes.

4. Squeeze one drop of sodium hydroxide onto a piece of each of the coloured indicator papers and record the colour you observe.

5. Pour one drop of each of the liquid indicators into each of the two test tubes of sodium hydroxide. Record the colour change you observe.

6. Wash out all the test tubes and droppers thoroughly.

7. Repeat steps 2 to 6 for aqueous ammonia and then for calcium hydroxide.

8. Compare the results you found for the various alkalis. Did all the alkalis give the same results for each indicator?

The word **alkali** is of Arabic origin, meaning dry. Alkalis belong to the set of bases but they are soluble in water, hence the name **hydroxide**. Dilute solutions of alkalis feel soapy and have a bitter taste. All alkalis conduct electricity. Alkalis can be identified by their colour changes with indicators.

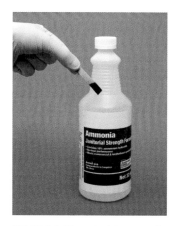

FIG 10.4.1 Household ammonia solution is strongly alkaline

Fun fact

A solution of ammonia gas dissolved in water used to be called ammonium hydroxide but it is now more commonly known as aqueous ammonia or ammonia solution.

Key terms

alkali a member of a group of substances that turn red litmus paper blue

hydroxide a compound of a metal with hydrogen and oxygen, which is often basic; if it is soluble it will form an alkaline solution

Check your understanding

1. Name three common alkalis.

2. What is the effect of an alkali on red litmus paper?

Strength of an alkali

We are learning how to:

- distinguish between substances that are acids and alkalis
- identify the strength of an alkali.

Strength of an alkali >>>

Activity 10.5.1

Exploring the strength of alkalis

Here is what you need:

- Universal indicator solution
- Small pieces of universal indicator paper
- Tweezers
- Spatulas
- Scalpels
- Experiment trays
- Test tubes
- Droppers
- Bleach
- Dishwashing liquid
- Baking soda
- Sodium
- Aqueous ammonia
- Calcium hydroxide.

SAFETY

Observe the safety icon on the alkali bottles. All indicator papers should be held with the tweezers. Avoid spillage. Take care when using scalpels.

Here is what you should do:

1. Copy Table 10.5.1. List the names of the substances you are testing, as shown below.

| Specimen | Universal indicator (colour change) | |
	Paper	Solution
bleach		
dishwashing liquid		
baking soda		

TABLE 10.5.1

2. Using a scalpel, spatula or dropper, place a tiny sample of each specimen in the cavities of the experiment tray.

3. Wash your hands and all the apparatus you used thoroughly, and dry your hands.

4. Place one little piece of indicator paper in each specimen. Observe and record the colour change.

5. Add one drop of indicator to each specimen. Observe and record the colour change.

6. Were all the colours the same?

7. Do you think that the difference in colour has any significance for an alkali's strength?

Universal indicator also shows the strength of alkalis. Universal indicator displays colours for alkalis with a pH between 8 and 14.

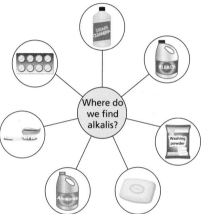

FIG 10.5.2 Alkalis around us

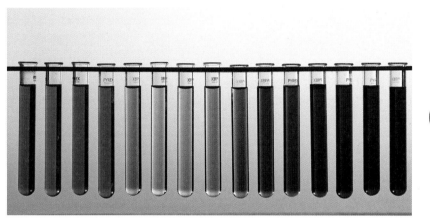

FIG 10.5.1 The colour moves from blue to dark purple as the alkali gets stronger

Alkalis can be found all around us, especially in the home.

So far you have identified acidic and alkaline substances. Based on the indicators, it can be seen that they are at opposite ends of the pH scale.

Check your understanding

1. What would the pH of a strong alkali be?

2. What numbers indicate the pH range of alkalis?

3. How does universal indicator show the strength of an acid?

4. Which number would indicate the strongest alkali?

Fun fact

An ant injects acid under your skin when it bites. By placing an ice cube on the bite, it is possible to soothe the sting and prevent swelling.

Making an acid–alkali indicator

We are learning how to:

- make an acid–alkali indicator.

Making an acid–alkali indicator ›››

You should already be familiar with some **acid–alkali indicators** from Units 10.2 and 10.3. Some indicators are obtained from organic material.

Litmus is extracted from lichen. This is a rather unusual organism in which algae and fungi co-exist for mutual benefit.

There are many organic materials that contain coloured chemicals which might provide acid–alkali indicators. For a chemical to be any use as an indicator it must be different colours in acids, and in alkalis.

Organic materials that contain coloured chemicals that may be useful as indicators include the petals of flowers like hibiscus, the leaves of vegetables like red cabbage and the roots of vegetables like beetroot.

Whatever the source, the coloured chemical must be extracted by physically crushing the organic material to break down the cells and then extracting it with a suitable solvent. Many of the chemicals responsible for colour are soluble in water but are more soluble in a solvent like ethanol. When extracting, it is important not to use too much solvent or the extract will be very dilute and colours will be difficult to see.

FIG 10.6.1 Litmus is obtained from lichen

FIG 10.6.2 Plant petals

FIG 10.6.3 Vegetable leaves

Activity 10.6.1

Testing for acids and alkalis

Here is what you need:

- Chopped up coloured plant material
- Mortar and pestle
- Beaker
- Ethanol
- Stirring rod
- Test tubes × 3
- Test tube rack

FIG 10.6.4 Vegetable roots

- Dropper pipette
- Dilute hydrochloric acid
- Dilute sodium hydroxide solution
- Eye protection.

Here is what you should do:

1. Place a small amount of chopped up coloured plant material in a mortar and pestle and grind it into a mushy paste.

2. Scrape it into a beaker.

3. Repeat this 3 or 4 times with small amounts of material each time.

4. Pour a small amount of ethanol into the mortar and wash the mortar and pestle to dissolve any coloured juice.

5. Pour the ethanol into the beaker with the crushed material and stir the mixture for a few minutes so that any coloured juice has a chance to dissolve in the ethanol.

6. Allow the contents of the beaker to settle for a few minutes.

7. Decant off the liquid from the beaker into a test tube, taking care not to pour out any of the plant material.

8. Pour dilute hydrochloric acid into a test tube to a depth of about 1 cm.

9. Add a few drops of your indicator solution using a pipette. Record the colour of the indicator in a table like the one below.

10. Repeat steps 8 and 9 using dilute sodium hydroxide solution in place of the acid.

| Colour of indicator in acid | |
| Colour of indicator in alkali | |

TABLE 10.6.1

> **Fun fact**
>
> It is thought that litmus was first used around 1300 by the Spanish alchemist Arnaldus de Villa Nova but it is unlikely that he used it as an acid–alkali indicator. Many coloured organic extracts like this were used in those days to dye cloth.

Check your understanding

1. Why is it essential that an indicator is different colours in acids and alkalis?

2. Why is it necessary to crush organic material before extracting any coloured material?

3. Why is ethanol preferred to water for extracting the coloured chemical from crushed organic material?

Key terms

acid–alkali indicator chemical that is different colours in acids and in alkalis

litmus common acid–alkali indicator

Acid–alkali reactions

We are learning how to:

• describe chemical reactions involving acids
• define neutralisation.

Acid–alkali reactions

Activity 10.7.1

Exploring neutralisation

For this experiment you need to work in two groups: one with an acid and the other without.

Here is what you need:

- A dilute acid
- A dilute alkali
- Liquid indicator (phenolphthalein)
- Measuring cylinder
- Droppers
- Test tube
- Straw.

 SAFETY

Observe the safety icon on the reagent bottles. Remember to observe safety rules when working with hazardous materials. Avoid spillage.

Here is what you should do:

Follow these instructions carefully and correctly.

1. Measure 1 cm³ of alkali and pour it into the test tube.

2. Use a dropper to place one drop of indicator into the alkali.

3. Observe and record the colour.

For the groups with the acid:

4. Using a second dropper, apply acid to the alkali in the test tube and shake it gently after each application.

5. Keep adding until a colour difference appears.

6. Use a third dropper to add drops of alkali to observe colour changes.

7. Add acid and alkali alternately until you think you have found a midpoint between acid and alkali.

8. Explain what you experienced as you changed from adding acid to alkali and back.

For the group with no acid:

9. Get one student to gently blow exhaled air through the straw into the alkali with the indicator.

10. They should keep exhaling until the colour changes.

11. Use the second dropper to add drops of alkali to observe colour changes.

12. Then the student should exhale into this combination.

13. Repeat until you think you have found a midpoint between exhaled air and alkali.

Alkalis are bases that are soluble in water. Acids and alkalis have opposite chemical properties. When an alkali and acid combine, there is a point where they both cancel out the effect of each other and this is called **neutralisation**. A neutral substance has a pH of 7, the midpoint of acids and alkalis.

Since acids and alkalis chemically react with each other, a new product is formed.

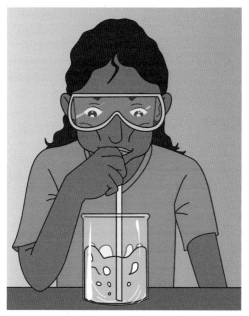

FIG 10.7.1 Carbon dioxide, which is slightly acidic, neutralises limewater, calcium hydroxide, which is an alkali

Key term

neutralisation the effect of acid and alkali cancelling each other out

Check your understanding

1. Describe how to explore the neutralisation of an acid by an alkali.

Neutralisation

We are learning how to:

- describe what happens in a neutralisation reaction.

Neutralisation >>>

The components of a chemical change occur in fixed or given proportions. To know the exact amount of each chemical to be used for neutralisation, you can use an indicator.

Fun fact

In ancient times, Roman soldiers were sometimes paid in salt (Latin: *sal*) – hence the origin of the word 'salary'.

Activity 10.8.1

Exploring neutralisation

This is a demonstration activity. You are to observe and record.

Here is what you need:

- Dilute sodium hydroxide
- Dilute hydrochloric acid
- Conical flask
- Methyl orange indicator
- Titration apparatus
- Evaporating dish
- Heat
- Tripod
- Gauze.

 SAFETY

Take care with chemicals and heat sources.

Here is what you should do:

1. Add one drop of indicator to 10 cm³ of sodium hydroxide, and mix in a conical flask.

2. Use the titration apparatus shown in Fig 10.8.1 to find and record the amount of hydrochloric acid used for neutralising alkali.

3. Use the data to combine a fresh set without the indicator.

4. Pour some of the combination into an evaporating dish and place over a Bunsen flame.

5. As the liquid evaporates, reduce the heat to a gentle flame.

6. If it's a sunny day, some of the combination may be placed outside on a crystallisation dish.

7. When evaporation is complete, examine the residue.

8. Write a laboratory report on this activity.

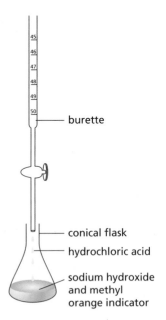

FIG 10.8.1 Titration apparatus set up to find neutralisation point

The product of the neutralisation of sodium hydroxide and hydrochloric acid is a **salt** and water. The evaporation occurring was that of the water. The residue is the salt (sodium chloride). The chemical reaction that occurred is shown thus:

sodium hydroxide + hydrochloric acid → sodium chloride + water

Products of neutralisation ⟫

The products of the neutralisation of an alkali and an acid are a salt and water.

The names of salts have two parts. The first part is the name of the metal involved in the reaction. (Ammonia is not a metal but when it reacts with acid, the first part of the name of the salt is ammonium.)

The second part comes from the acid used.

If the acid is

- hydro**chlor**ic acid then the salt is a **chlor**ide

- **nitr**ic acid then the salt is a **nitr**ate

- **sulf**uric acid then the salt is a **sulf**ate.

FIG 10.8.2 There are many different metal salts and some of them are brightly coloured

Check your understanding

Write the products of the following neutralisations:

1. magnesium hydroxide + hydrochloric acid

2. aluminium hydroxide + hydrochloric acid

3. copper hydroxide + hydrochloric acid

4. zinc hydroxide + hydrochloric acid

5. lead hydroxide + hydrochloric acid

6. aqueous ammonia + hydrochloric acid

7. aluminium hydroxide + nitric acid

8. sodium hydroxide + sulfuric acid

9. copper hydroxide + nitric acid

10. magnesium hydroxide + sulfuric acid

Key terms

titration a technique in which the concentration of one solution is found by using another solution with a known concentration

residue the material remaining after distillation, evaporation, or filtration

salt the product of neutralisation

Neutralisation reactions in everyday life

We are learning how to:

- identify neutralisation reactions that take place in our everyday lives.

Neutralisation reactions in everyday life ⟫

Chemical reactions are not limited to test tubes in the laboratory. There are many examples of chemical reactions taking place in our everyday lives, including neutralisation reactions. Some have already been mentioned in this unit.

Baking powder ⟫

Bread and cakes often have a light texture so they are easy to bite into and chew. Bakers achieve this by adding **baking powder** to the dough before baking. This causes the dough to rise.

Baking powder contains an acid and a carbonate or a hydrogencarbonate. These chemicals react to produce **carbon dioxide** gas (you will learn more about this reaction in lesson 10.11). Here is a typical example of the chemical reaction that takes place:

tartaric acid + sodium hydrogencarbonate →
sodium tartrate + carbon dioxide

The bubbles of carbon dioxide are contained within the dough and expand when the dough is heated in the oven. The result is lots of 'holes' in the bread or cake, giving it a light texture.

Antacids ⟫

The lining of the stomach releases hydrochloric acid to assist in the process of digestion. The more we eat, the more acid is produced.

When the stomach produces lots of acid we get stomach pain, which we call indigestion. In order to reduce the pain we might take an '**antacid**', which neutralises the excess acid.

There are lots of different brands of antacids available. They contain carbonates, hydrogencarbonates or weak hydroxides like magnesium hydroxide. All of these compounds will neutralise the stomach acid.

FIG 10.9.1 Baking powder contains an acid and a carbonate/hydrogencarbonate

FIG 10.9.2 Bread with a light texture

FIG 10.9.3 Sometimes we eat too much

FIG 10.9.4 Antacids neutralise excess stomach acid

Have you ever eaten sweets like a sherbet that fizz and tingle on your tongue?

These sweets contain an acid and a carbonate/hydrogencarbonate which are safe to eat. The powder doesn't do anything when it is dry but as soon as you drop it on your tongue a chemical reaction starts which produces carbon dioxide. This chemical reaction makes your tongue tingle!

FIG 10.9.5 Sherbet fizz

Activity 10.9.1

Home-made fizzy sherbet

You should carry out this activity in the kitchen at home. Your teacher might set this for homework and test the fizzy sherbet you have made.

Here is what you need:

- 1 measure of sodium hydrogencarbonate (this is often called sodium bicarbonate or baking soda)

- 1 measure of citric acid (this is the acid from citrus fruits like lemons).

The size of the measure will depend on how much you plan to make. You should make small amounts until you hit on a mixture that tastes really good.

You can also add substances that will modify the taste and appearance of your sherbet. For example:

- icing sugar will make it taste sweeter

- jelly crystals will give it some colour.

You need to experiment with the amounts you add to produce the perfect fizzy sherbet!

Here is what you should do:

1. Mix the ingredients as powders in a bowl. Don't add any water or your mixture will react and lose its fizz.

2. Keep your fizzy sherbet dry in a suitable container like a small food box.

3. Test it out on your family and friends.

Check your understanding

1. Milk of magnesia contains magnesium hydroxide. How is it able to relieve stomach ache?

Key terms

baking powder mixture of chemicals used to make a dough rise before baking

carbon dioxide gas produced by the reaction of an acid with a carbonate or hydrogencarbonate

antacid medicine that relieves stomach ache by neutralising excess stomach acid

Ammonium compounds

We are learning how to:

- make and test for ammonia gas
- use neutralisation reactions to make ammonium compounds.

Ammonia gas »»

Ammonia is a **pungent** gas that must be handled with great care. It will dissolve in the moisture on the eyeballs or in the nose and produce an alkaline solution that will cause irritation.

When ammonia dissolves in water a weak alkali is formed. You have already tested aqueous ammonia with **acid–alkali indicators**.

This solution is often called ammonium hydroxide but the term 'aqueous ammonia' is a more accurate description. Although the chemical formula is often given as NH_4OH, it is better represented as $NH_3 \cdot H_2O$.

Ammonia gas can be made by heating any ammonium compound with an alkali. The combination of compounds most often used is shown in the following equation:

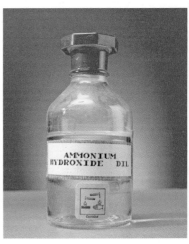

FIG 10.10.1 Ammonia solution is often called ammonium hydroxide

ammonium chloride + calcium hydroxide →
ammonia + calcium chloride + water

Activity 10.10.1

Making ammonia gas

Here is what you need:

- Ammonium chloride
- Calcium hydroxide
- Boiling tube
- Delivery tube
- Stand and clamp

- Tweezers
- Red litmus paper
- Test tube
- 250 cm³ beaker.

Here is what you should do:

1. Mix equal amounts of solid ammonium chloride and calcium hydroxide and place the mixture in a boiling tube. Only use a small sample.

2. Support the boiling tube horizontally using a stand and clamp.

Fun fact

The maximum concentration of ammonia it is possible to have in water has a **density** of 0.880 g/cm³. It is often referred to as '880 ammonia'.

3. Place a bung and delivery tube as shown in Fig 10.10.2.

4. Gently heat the mixture until you are aware of a pungent smell.

5. Place an inverted test tube over the delivery tube and collect a test tube of ammonia gas.

6. Hold a piece of red litmus paper with tweezers, dampen it and place it into the bottom of the test tube. What colour does it turn?

7. Two-thirds fill a beaker with water.

8. Keeping the test tube inverted, push the open end into the water in the beaker and observe what happens.

9. Explain your observations.

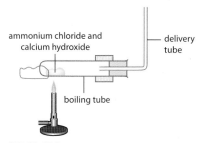

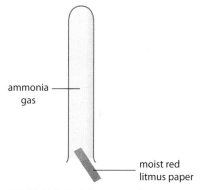

FIG 10.10.2

FIG 10.10.3 Test for ammonia

Ammonia is the only common gas that is alkaline, so turning damp litmus paper from red to blue is a test for the presence of this gas.

Ammonia is less **dense** than air and is very soluble in water.

Ammonium compounds 》》

Ammonia forms salts by neutralising acids. These salts are ammonium compounds.

Ammonium nitrate and ammonium sulfate are made in large quantities and used by farmers to increase the fertility of their soils.

Ammonium compounds provide plants with extra nitrogen so they grow well and produce large crops.

FIG 10.10.4 Ammonium compounds are an important source of nitrogen for plants

Check your understanding

1. Would dry ammonia gas turn a piece of dry red litmus paper blue? Explain your answer.

2. Phosphoric acid is another strong acid.

   a) What is the name of the compound formed when ammonia neutralises phosphoric acid?

   b) Write a word equation for this reaction.

Key terms

pungent having a sharp smell

acid–alkali indicator chemical that turns different colours in acidic and alkaline conditions

density/dense mass per unit volume

Reactions of acids

We are learning how to:

- describe the reaction between acids and carbonates
- describe the reaction between acids and metals.

Acid–carbonate reactions 》》

Activity 10.11.1

What gas is given off during an acid–carbonate reaction?

This is a demonstration lesson. Volunteers will be needed. Observe carefully.

Here is what you need:

- Dilute hydrochloric acid
- Samples of metal carbonates
- Test tubes
- Splint
- Bunsen burner
- Limewater solution
- Straw.

 SAFETY
Observe care with acids.

Here is what you should do:

1. Place a little carbonate in a test tube.
2. Have a lighted splint available.
3. Pour some acid onto the carbonate and place the splint into the evolving gas. What happens to the flame?
4. Pour some colourless limewater solution $Ca(OH)_2$ into a test tube.
5. Allow the evolving gas to flow into the test tube of limewater and shake it.
6. What has happened to the limewater? Can you identify the gas?
7. Using the straw, exhale into a test tube of limewater. What do you observe?
8. What gas was exhaled?
9. Can you now identify the gas that was evolved from the acid–carbonate reaction?

When an acid reacts with a **carbonate** or a hydrogencarbonate, a gas is produced (evolved). To identify the gas, it must be tested. When a lighted splint was brought into the gas it was extinguished. When the gas was tested with limewater, a white precipitate resulted. Carbon dioxide is the only gas that forms a white precipitate with limewater. The gas evolved from an acid–carbonate reaction is, therefore, carbon dioxide.

FIG 10.11.1 Acid–carbonate reactions:
a) copper carbonate **b)** sodium carbonate

Activity 10.11.2

What happens when acids react with metals?

Here is what you need:

- Rack of labelled test tubes, each containing a metal: aluminium, magnesium, copper, zinc, lead
- An acid.

 SAFETY
Observe the name and the safety icon on the acid bottle.

Here is what you should do:

1. Take out each metal, observe it and replace it in the test tube.

2. Pour some acid onto each metal, shake gently, then feel the outside of the test tube. Look for any reaction and record any observations you make.

3. You can report your findings as a group. Did each metal react the same way to the different acids?

4. With which ones was there effervescence? What does effervescence indicate?

FIG 10.11.2 Reaction between zinc and acid

The gas evolved from the reaction of acids and metals is hydrogen.

When a lighted splint comes into contact with hydrogen in the air, it produces a squeaky pop. This is because hydrogen is an explosive gas. Hydrogen is the only gas that reacts that way with a flame.

> **Fun fact**
>
> Hydrogen is the least dense gas. In the past, hot air balloons were filled with hydrogen. However, because of the risk of explosion the use of hydrogen was discontinued and helium is now used to fill the balloons.

Check your understanding

1. Describe the reaction of an acid with a carbonate.

2. Describe the reaction of an acid with a metal.

3. Copy and complete these equations:

    a) calcium + sulfuric acid →

    b) lead + nitric acid →

    c) aluminium + sulfuric acid →

    d) zinc + hydrochloric acid →

> **Key term**
>
> **carbonate** substance containing a metal, carbon and oxygen

Safety booklet

We are learning how to:

- write a booklet providing advice to keep people safe when working in the laboratory.

Safety ⟩⟩⟩

Safety is not an issue which is confined to the chemistry laboratory. A typical kitchen contains sharp knives, hot pans and fruit juices that sting the eyes.

We are aware of these dangers from an early age and we learn how to keep ourselves safe. The school laboratory is no more dangerous than your kitchen but some of the hazards are different.

FIG 10.12.1 A kitchen is a potentially dangerous place

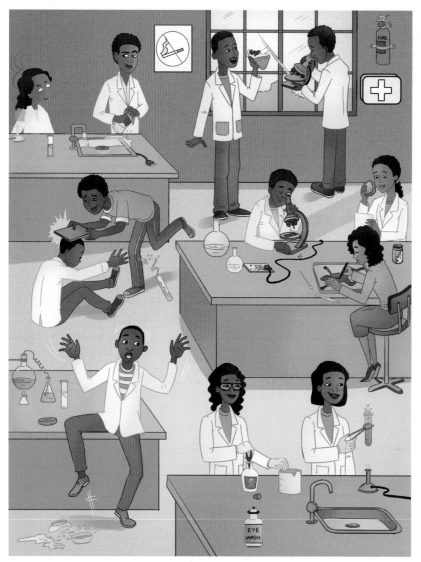

FIG 10.12.2 Foolish behaviour leads to accidents

Your teacher will not ask you to carry out experiments which are dangerous but there are always hazards when dealing with chemicals so it is important to take precautions and behave correctly.

Whenever you carry out experiments involving chemicals like acids and alkalis you must wear eye protection. Splashes of acids and alkalis can irritate the eyes and cause permanent damage.

Your laboratory should be equipped with an eye wash bottle just in case somebody does get chemicals in their eyes.

Even dilute acids and alkalis are **corrosive** and will damage skin. Any spillages should be washed off the skin immediately with cold water.

These chemicals will also damage the fabrics from which your clothing is made. This may not show up as holes straight away, but the next time your clothes are washed they will come out in holes. If you splash acid or alkali onto your clothing, wash the area with lots of cold water immediately. The area in which you work should always be kept clean and tidy. Any spillages should be mopped up straight away with a wet cloth.

FIG 10.12.3 Protect your eyes

FIG 10.12.4 Eye wash bottle

Activity 10.12.1

Producing a safety booklet

Your task is to produce a booklet about safety using acids and alkalis in the laboratory. This will be four pages which you will make by folding a sheet of letter-size paper in half along the short side.

It is up to you to decide on the format. It might:

- have a mixture of text and diagrams
- take some photographs to illustrate your text
- make use of tables and/or bullet points
- use colour to emphasise important points.

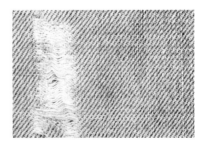

FIG 10.12.5 Acids and alkalis are corrosive. This denim cloth has been damaged by acid.

Check your understanding

State why each of the following actions is potentially dangerous in the chemistry laboratory.

1. Leaving your school bag on the floor between the benches.

2. Eating a snack while you are handling chemicals.

3. Looking down into a boiling tube while you are heating it.

4. Putting hot glassware directly onto the bench.

Key term
....

Corrosive destroys by chemical action

Review of Acids and alkalis

- Acids are chemical substances that have a sour taste.

- Alkalis are chemical substances that conduct electricity and have a soapy feel when diluted.

- Both acids and alkalis are identified by the use of indicators made from special dyes.

- There are a variety of indicators used for identifying acidity or alkalinity.

- Universal indicator gives the degree of acidity or alkalinity using a range of colours numbered from 0 to 14.

- The midpoint on the universal indicator is green and indicates neutrality (pH7).

- Carbon dioxide is a gas that can form a weak acid.

- Bases form a set of chemicals that include oxides, hydroxides and carbonates.

- Acids–alkali indicators can be made from coloured organic material.

- Acids and bases can neutralise each other to form new products.

- Acids also react with carbonates and metals to form neutral products called salts.

- CO_2 changes calcium hydroxide (limewater) to a white precipitate.

- H_2 explodes, producing a squeaky pop.

- A salt is named after the reactants from which it was formed.

- Hydrochloric acid forms chlorides.

- Nitric acid forms nitrates.

- Sulfuric acid forms sulfates.

- Neutralisation reactions in everyday life include using baking powder to make bread and cakes, using antacids to relieve indigestion, and fizzy sweets.

- Ammonia gas turns damp red litmus paper blue.

- Ammonia is the only common alkaline gas.

- Ammonia gas is less dense than air and very soluble in water.

- Ammonia forms salts by neutralising acids.

- It is essential to create a safe environment when carrying out experiments.

- Eye protection must be worn when working with potentially harmful materials.

- Even dilute acids and alkalis are corrosive.

Review questions on Acids and alkalis

1. You have found a substance. Explain and give the results of three sets of tests you would carry out to discover whether the substance is acidic or alkaline.

2. Explain what you would do to determine how strong an acid or an alkali is.

3. Briefly describe how you would make an acid–alkali indicator from coloured flower petals.

4. When an acid neutralises a base, are the products acidic or alkaline?

5. Outline the steps you would follow to show that the result of neutralisation is a salt.

6. Complete the sentences by filling in the blanks below.

   Salts made from

   a) hydrochloric acid are _____.
   b) nitric acid are _____.
   c) sulfuric acid are _____.

7. During a reaction there is vigorous effervescence. Describe what you would do to conclude whether the gas evolved was carbon dioxide or hydrogen.

8. Show four different ways in which you can produce zinc nitrate.

9. Copy and complete the following chemical equations:

   a) zinc hydroxide + sulfuric acid →
   b) magnesium carbonate + nitric acid →
   c) aluminium hydroxide + sulfuric acid →
   d) copper oxide + nitric acid →
   e) lead + sulfuric acid →
   f) sodium + hydrochloric acid →

10. **a)** Explain why any spillage of an acid onto the bench should be mopped up immediately with a damp cloth.

    b) Some acid has been spilt on the floor. Name a chemical that could be sprinkled as a solid onto the floor to neutralise the acid.

    c) The student next to you has splashed some acid into her eye. What action should be taken?

11. **a)** Write a word equation for the formation of ammonia gas from ammonium chloride and calcium hydroxide.

    b) Explain why ammonia gas:

       i) is collected from a delivery tube pointing upwards

       ii) cannot be collected over water.

    c) Describe a test to show that an unknown gas is ammonia.

Using citric acid in dyeing

Natural dyes from plants, insects and minerals have been used for thousands of years to colour natural textiles like cotton and wool. Sometimes substances like citric acid are added to the dye bath in order to alter the pH of the solution. This helps the dye bond to the fabric.

A local eco-friendly fabric company has decided to produce a new range of colours for woollen fabrics using only natural products. They want you to use your knowledge of acids and bases to test the effectiveness of citric acid, obtained from lemons or limes, for fixing the following natural dyes.

FIG 10.SIP.1 Wool fabrics

Natural dye source	Colour
Plum skins	Pink
Turmeric	Yellow
Spinach or sorrel	Green
Beets	Red-purple
Coffee or tea	Brown

1. You are going to work in groups of 3 or 4 to investigate if citric acid helps to fix the range of dyes. Your teacher will advise you on how many dyes to investigate according to how much time is available. The tasks are:

   - To review the pH scale and the effect that acids have on the pH value of a solution.

   - To obtain solutions of natural dyes.

   - To obtain a supply of citric acid from lemons or limes.

   - To obtain some undyed woollen strands or fabric.

   - To test the effect of citric acid in the dye mixture.

   - To consider how the dyeing procedure may be modified.

   - To write a report, including a PowerPoint presentation in which you describe what you did and what you found out. This should include wool samples to illustrate your results.

   a) Look back through the unit and make sure you are familiar with the pH scale and what happens to the pH of a solution when an acid is added to it.

   b) You need to devise a method of making the natural dyes that you need to test. A simple way to do this is to boil the dye source in water for a short time, then filter to obtain a dye solution.

   c) You need to devise a method of extracting citric acid from lemons or limes. There are lots of simple devices available.

FIG 10.SIP.2 Devices for extracting juice from lemons

You need to have a stock solution of citric acid.

d) You need to develop a system which you can use on each natural dye. For example:

i) Divide the dye solution into three parts and label them A, B and C.

ii) Measure the pH of solution A.

iii) Add lemon juice to solution B to reduce its pH by 1 or 2.

iv) Add lemon juice to solution C to reduce its pH by 3 or 4.

v) Label wool samples A, B and C.

vi) Place samples of wool into solutions A, B and C for the same amount of time.

vii) Remove the wool samples and allow them to dry.

viii) Compare the colour of the samples.

ix) Cut the wool samples in half.

x) Wash one half of each sample in clean water.

xi) Allow the samples to dry and compare the colours before and after washing.

Take photographs as you work, particularly of the dyed wool samples before and after washing so they can be compared.

e) After carrying out your initial tests, think about how you might extend your research. Here are some ideas.

- Does temperature affect how well a dye fixes?

- Does time affect how well a dye fixes?

- Does lowering the pH by 5 or 6 give different results?

- Does washing with soapy water remove more dye than washing with pure water?

f) Prepare a PowerPoint presentation that describes what you did in order to test how well the natural dyes were fixed at different pH values. Show samples and photographs to illustrate your findings.

Unit 11: Embryo development and birth control

We are learning how to:

- identify the parts of the female reproductive system associated with fertilisation
- describe what happens during the process of fertilisation.

Fertilisation ⟫

Fertilisation is the joining or fusion of a male sex cell or gamete and a female sex cell or gamete. In human reproduction the sperm is the male sex cell and the egg is the female sex cell. **Fertilisation** takes place in the body of the female.

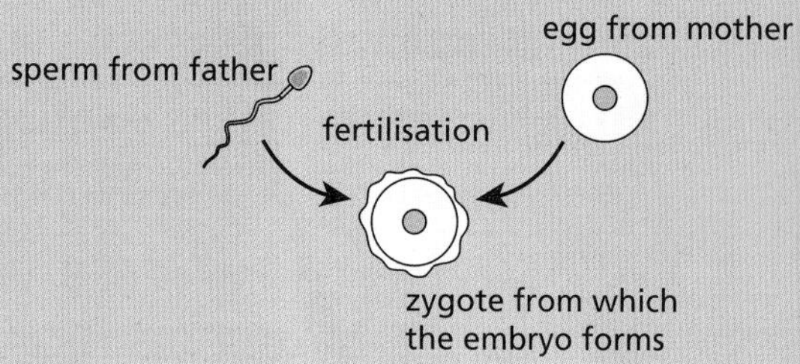

During fertilisation the sperm and the egg combine to form a **zygote**. It is from the zygote that the embryo will form and develop to become a new-born baby.

An egg cell is only fertilised by a single sperm cell. Once a sperm has entered the egg rapid changes take place to the membrane surrounding the egg. As a result the membrane prevents any additional sperm cells from entering.

sperm from father

fertilisation

egg from mother

zygote from which the embryo forms

FIG 11.1.1 Fertilisation

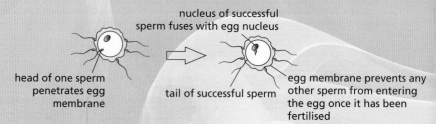

nucleus of successful sperm fuses with egg nucleus

head of one sperm penetrates egg membrane

tail of successful sperm

egg membrane prevents any other sperm from entering the egg once it has been fertilised

FIG 11.1.2 An egg is only fertilised by one sperm

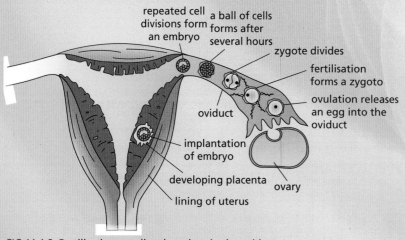

repeated cell divisions form an embryo

a ball of cells forms after several hours

zygote divides

fertilisation forms a zygoto

ovulation releases an egg into the oviduct

oviduct

implantation of embryo

developing placenta

lining of uterus

ovary

FIG 11.1.3 Fertilisation usually takes place in the oviduct

Fun fact

The oviducts are also called the Fallopian tubes.

Fertilisation usually takes place in the woman's oviduct as the egg is passing down from the ovary. Soon after the zygote forms it starts to divide and this process continues as more and more cells are formed. At this time the zygote becomes an **embryo**.

In this part of the woman's menstrual cycle, her uterus wall has become thickened ready to receive an embryo. As the embryo enters the uterus it sinks into the wall in a process called implantation.

Finger-like projections called villi extend from the embryo into the uterus wall. The villi bind to the uterus wall, forming a placenta which joins the embryo to its mother. You will learn more about this in the next lesson.

Over the next couple of weeks the attachment develops into the umbilical cord. By the ninth week of pregnancy the major body organs are present in the embryo and it is recognisable as human. At this time the embryo is described as a **foetus**. This term is used for the developing baby until its birth.

> **Fun fact**
>
> Foetus is often also spelt as 'fetus'. You will see both terms used on the Internet and in books. They mean the same thing.

Activity 11.1.1

Identifying the different parts of the female reproduction system

Carry out this activity on the school field.

Here is what you need:

- String
- Pegs
- Bamboo sticks to hold labels
- Sticky tape
- Card for labels.

1. Use pegs and string to mark out an outline of the female reproductive system. Use Fig 11.1.3 to guide you.

2. Walk through your outline and place labels where important things happen, such as:

   a) Egg released

   b) Fertilisation takes place

   c) Zygote starts to divide

   d) Embryo embeds.

3. Take some photographs of your model or make a short role play video.

Check your understanding

1. Where does fertilisation normally take place?

2. When does an embryo become a foetus?

3. Why is it than an egg can only be fertilised by one sperm?

4. What happens to the zygote shortly after it is formed?

> **Key terms**
>
> **fertilisation** coming together of the male and female sex cells
>
> **zygote** formed by the combination of a male sex cell and a female sex cell
>
> **embryo** develops from a zygote
>
> **foetus** the embryo at about 9 weeks of pregnancy when recognisable as human

The role of the placenta

We are learning how to:

- describe the structure of the placenta
- explain the role of the placenta during foetal development.

The placenta »»

The **placenta** is a flattened circular organ that develops on the wall of the uterus. The **umbilical cord** develops with the foetus and connects the foetus to the mother at the placenta. It is literally a lifeline for the foetus.

Everything the developing baby needs throughout the nine months of pregnancy passes from the mother along the umbilical cord. Waste products produced by the developing baby pass in the opposite direction.

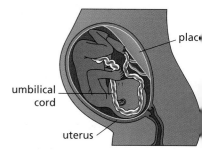

FIG 11.2.1 The umbilical cord and placenta

labels: plac(enta), umbilical cord, uterus

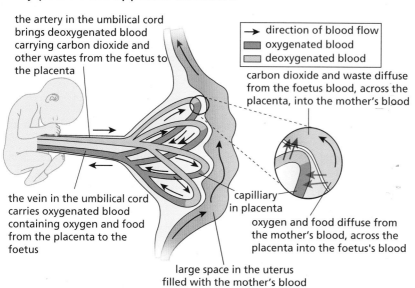

the artery in the umbilical cord brings deoxygenated blood carrying carbon dioxide and other wastes from the foetus to the placenta

→ direction of blood flow
oxygenated blood
deoxygenated blood

carbon dioxide and waste diffuse from the foetus blood, across the placenta, into the mother's blood

the vein in the umbilical cord carries oxygenated blood containing oxygen and food from the placenta to the foetus

capilliary in placenta

oxygen and food diffuse from the mother's blood, across the placenta into the foetus's blood

large space in the uterus filled with the mother's blood

FIG 11.2.2 Materials pass along the umbilical cord

Blood passes from the placenta to the embryo along a vein and blood returns from the embryo to the placenta through an artery.

Although the nutrients the embryo requires are provided from the mother's blood supply, the circulatory system of the embryo cannot be directly connected to the mother because the high blood pressure in the mother's arteries would burst the tiny blood vessels in the embryo. Instead substances pass between the blood of the embryo and that of the mother in the placenta by a process called **diffusion**.

Oxygen and food, in the form of glucose and other substances, diffuses from the mother's blood across the placenta into the blood of the embryo. At the same time waste products like carbon dioxide and urea pass in the opposite direction.

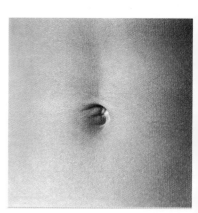

FIG 11.2.3 Navel or tummy button

FIG 11.2.4 Placenta and part of umbilical cord

The fluid-filled sac acts as a shock absorber which protects the embryo or foetus from bangs and bumps.

By 10 weeks the embryo has grown to about 6 cm and is now called a foetus. This might not sound very big, but between 6 and 10 weeks the embryo has grown from 12 mm to 60 mm or 5 times its earlier size.

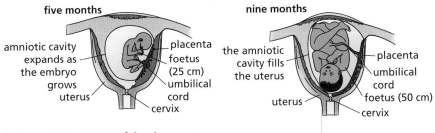

FIG 11.3.3 Later stages of development

The foetus is nourished by the mother via the umbilical cord throughout pregnancy. As the foetus develops and grows in size the amniotic cavity grows bigger until, when close to birth, it fills the uterus. The cervix supports the foetus as it grows and prevents it falling out of the uterus.

When the 'waters break' this means that the amniotic sac has ruptured. It is a sign that the baby is about to be born.

Activity 11.3.1

Creating a booklet to show the stages of growth of the foetus

You will work in a group for this activity.

The text above describes the different stages in the development of the embryo and foetus during pregnancy. Carry out further research about this and gather additional information in the form of text, images and data. You might have a friend or relative who is pregnant. What does it feel like? At what point in pregnancy does she say it feels like she is carrying another little human being?

Use the information you have to create an album or booklet about how the embryo and foetus develop.

Check your understanding

1. What is the normal gestation period in humans?
2. What does the 'waters breaking' usually indicate?
3. What is the function of the amniotic fluid?
4. About how long is the developing foetus after 5 months of the pregnancy?
5. What is the difference between cell division and cell differentiation?

Key terms

cell division replication to increase the number of cells

gestation period the time between conception and birth

amnion membrane that surrounds the foetus as it develops

amniotic fluid watery fluid contained in the amnion

Pre-natal care and maternal behaviour during pregnancy

We are learning how to:

- describe pre-natal care a mother may obtain during pregnancy
- understand that an expectant mother and father can make appropriate lifestyle choices to care for their unborn child.

Pre-natal care »»

Pre-natal care or antenatal care is the care provided for the mother and her developing baby before birth.

Expectant mothers regularly attend clinics to make sure they are in good health and learn about childbirth. Health centres may hold pre-natal classes, where expectant mothers can share their experiences. First-time mothers learn from women who have already given birth.

During pregnancy **ultrasound** scans are usually carried out on the mother. This uses very high-pitched sound waves to create an image of the foetus in the mother's uterus. Doctors can take measurements of the foetus's brain and spine from the scan to check that the foetus is developing properly.

In addition to the pre-natal care provided by health centres and doctors, the expectant mother must provide her own pre-natal care by ensuring that her lifestyle does not jeopardise the development of her unborn baby.

For example, there is scientific evidence that mothers who smoke during pregnancy tend to have smaller babies because chemicals entering the woman's blood during smoking will eventually find their way into the foetus' blood. Excessive consumption of alcohol and taking recreational drugs are also detrimental to the development of the baby.

Although it is the mother who carries the baby in many families the father supports his partner by giving up smoking and reducing his alcohol intake. In this way the mother and father each work towards having a healthy baby.

It is also possible for viruses to pass from the mother to the foetus during pregnancy. Some viruses, like **rubella** which is responsible for German measles, can cause the baby to be born with defects such as blindness. In some parts of the world, on reaching puberty, girls are vaccinated against German measles or rubella to ensure they are not affected by this virus in later life during pregnancy.

FIG 11.4.1 A pre-natal class

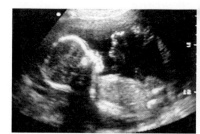

FIG 11.4.2 Ultrasound scan of a developing foetus

FIG 11.4.3 Smoking during pregnancy reduces birth mass

Another virus that may be passed on to an unborn baby by the mother is **HIV**. It is estimated that around the world over 1000 babies infected with HIV are born every day. The only way this can be reduced is by safer sexual practice reducing the number of HIV sufferers.

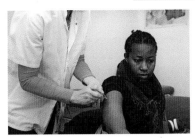

FIG 11.4.4 Vaccination against rubella

Activity 11.4.1

Drawing a bar chart to show data about birth mass and cigarette smoking

Here is what you need:

- Sheet of graph paper

- 30 cm ruler.

Table 11.4.1 shows the percentage of babies of different masses born to mothers who smoke, and those that don't smoke.

Mass at birth/kg	Percentage born to mothers who smoke	Percentage born to mothers who do not smoke
2.0–2.5	48	52
>2.5–3.0	38	62
>3.0–3.5	33	67
>3.5–4.0	26	74
>4.0–4.5	21	79
>4.5	17	84

TABLE 11.4.1

FIG 11.4.5 Babies may be born with HIV

Draw a bar chart to show this information. For each range of birth mass values you should draw two bars alongside each other; one bar to represent mothers who smoke and one bar to represent mothers who don't smoke.

Check your understanding

1. What does 'pre-natal' mean?

2. Which illness is caused by the rubella virus?

3. How is it possible to 'see' the foetus as it develops inside the mother?

4. How can the lifestyle of the mother affect the development of the foetus?

Key terms

pre-natal care the care a mother receives before giving birth while she is pregnant

ultrasound very high-frequency sound

rubella virus that causes German measles and which might damage an unborn baby

HIV human immunodeficiency virus which attacks the immune system

Contraception (1)

We are learning how to:

- describe different methods of contraception to prevent pregnancy.

Contraception ▶▶▶

Partners often wish to have sex but they don't want this to result in the woman becoming pregnant. To prevent pregnancy, different forms of contraception may be used.

Contraception allows people to choose when to have children and to determine the size of their family. This is called **birth control** or family planning. The decisions are personal and require a mature and responsible attitude. Women do not have to become pregnant by accident.

Methods of contraception can be divided into four groups. In this lesson we will look at examples of each group.

Natural methods ▶▶▶

These methods require modifications to our behaviour.

Requires self-discipline	How it works	Advantages	Disadvantages
Withdrawal or coitus interruptus	The male withdraws his penis before ejaculation	No side effects	Very unreliable as fluid secreted before ejaculation may contain sperm. Provides no protection from STIs
Rhythm	Intercourse is avoided at times when the woman is most likely to conceive	No side effects	Very unreliable
Abstinence	Refrain from sexual intercourse	No worries	Requires self-discipline

TABLE 11.5.1

Barrier methods ▶▶▶

These methods prevent the sperm from reaching the egg so that fertilisation cannot take place.

The effectiveness of barrier methods may be increased if they are used in conjunction with spermicidal creams.

Method	How it works	Advantages	Disadvantages
Condom	The condom is slid over the erect penis penis condom FIG 11.5.1	No side effects Protects against STIs Cheap to buy and readily available	Can only be used once Thought by some to reduce sensitivity
Coil / loop	Inserted into the uterus by a doctor and prevents implantation of a fertilised egg	Reliable and inexpensive Remains effective for several years	May interfere with menstrual cycle Can cause uterine infections
Cap / diaphragm	The cap is inserted and placed at the top of the cervix cap fitted over the cervix FIG 11.5.2	No side effects Offers some protection against STIs	May be damaged during intercourse

TABLE 11.5.2

Activity 11.5.1

You should work in a group for this activity. In some countries condoms are made available free of charge in places such as washrooms and hotel rooms.

- Those in favour of this say that this practice reduces unwanted pregnancies and the spread of STIs.
- Those against this say that this practice encourages sexual promiscuity.

Debate in your group whether you think this is a good practice or not.

Check your understanding

1. Why is 'coitus interruptus' not a reliable method of birth control?
2. Apart from contraception, what other advantage is there in using condoms?
3. How does a coil prevent pregnancy?

Key terms

birth control controlling if and when you have children, and the size of your family

condom latex sheath slid over the erect penis before intercourse

cup/diaphragm device inserted in the top of the cervix

Contraception (2)

We are learning how to:

- understand different methods of contraception.

Hormonal methods and spermicidal creams

The release of eggs each month in the female is controlled in the body by chemicals called hormones. **Hormonal methods** introduce hormones to interrupt this cycle.

Method	How it works	Advantages	Disadvantages
Hormone pill	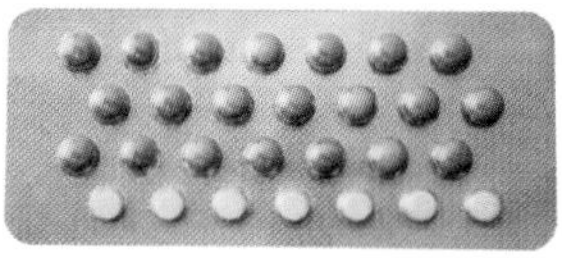 FIG 11.6.1 One taken each day prevents the body releasing eggs	Simple and reliable	Increases the risk of heart disease and high blood pressure
Spermicidal creams	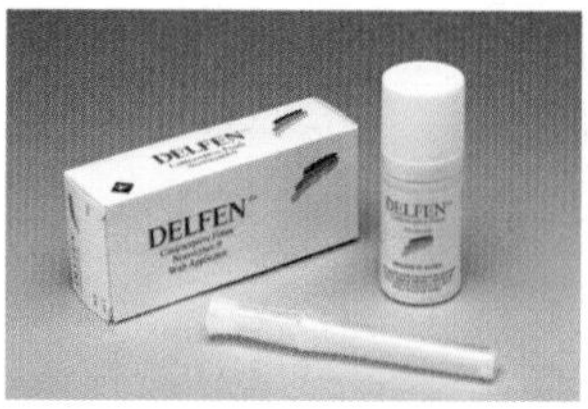 FIG 11.6.2 Contains chemicals that kill and block sperm from entering the uterus	Easy to obtain and simple to use	Not reliable but useful when used with some other methods of contraception

TABLE 11.6.1 Hormonal methods and spermicidal creams

Surgical methods

Surgical methods use simple procedures to modify the reproductive organs so that fertilisation cannot take place.

Family planning clinics are places where couples can go and get advice from doctors and other professional people on different forms of contraception, and on family planning.

Fun fact

The 'morning after' pill is an emergency contraceptive which is taken within two days of unplanned or unsafe sexual intercourse. It contains a hormone that prevents the implantation of an embryo.

Method	How it works	Advantages	Disadvantages
Vasectomy	The sperm ducts are cut and the ends tied off so sperm cannot travel from the testes to the penis sperm duct ends cut and tied epididymis testis FIG 11.6.3	100% effective No protection against STIs	Difficult to reverse
Tubal ligation	The oviducts are cut and the ends tied off so the eggs cannot travel down to the uterus bottom end of oviduct top end of oviduct ovary uterus vagina FIG 11.6.4	100% effective No protection against STIs	Difficult to reverse

TABLE 11.6.2

Activity 11.6.1

Researching the Jamaica National Family Planning Board

You should work in a group for this activity.

Find out all that you can about the Jamaica National Family Planning Board and the work that it does.

If possible, visit your local family planning clinic and ask about the work that they do. Invite someone from the clinic to visit your class to talk about the importance of family planning.

FIG 11.6.5 Logo of the Jamaica National Family Planning Board

Key terms

hormonal methods chemicals, most commonly in the form of a pill, containing hormones that prevent pregnancy

spermicidal killing sperm

vasectomy an operation to cut the sperm ducts to prevent the passage of sperm

tubal ligation an operation to cut the oviducts to prevent the passage of eggs

Check your understanding

1. a) What is a vasectomy?

   b) Why might some men think that this affects their virility?

2. A woman comes from a family that has a history of high blood pressure. What form of contraception should she avoid?

Teenage pregnancy

We are learning how to:

- describe the problems resulting from teenage pregnancy
- explain that pregnancy is more appropriate for a mature woman in a steady relationship.

Teenage pregnancy

Although the female body is able to have babies as soon as periods start, which can be as early as 11 or 12 years old, this does not mean that it is sensible for a girl to have a baby when she is young.

There is a lot more to having and successfully bringing up a baby than the biological process of having one. In this lesson we will discuss why it may not be a good idea for a girl still in her teens to have a baby.

A woman may not attain her adult size until she is late in her teens. Although the female body can conceive and a girl can become pregnant earlier than this, her body may not be strong enough to carry a baby. Pregnancy might cause her body internal damage which could affect her ability to have more babies later in life.

Bringing up a baby is an expensive enterprise. When a baby is born to a mature woman in a permanent relationship it is likely that she has some financial stability and a partner who can continue working to provide for his family. A teenage girl will have no such financial stability and is not likely to have a partner who can provide for her and her baby.

Bringing up a baby requires both knowledge and patience. A mature woman has had time to complete her education and has therefore learnt about pre- and post-natal care. She may also benefit from advice given her by older relatives and by watching how they raise their own babies. A teenage girl is not likely to have completed her education and may not be old enough to have learnt from others.

Having a baby when she is an adult means that a girl has sufficient time to complete her education and work at a job where she will acquire skills. Once a woman has these skills she will always have the option of returning to work some time in the future when her children are older. A girl who has a baby when she is young has not completed her education or acquired skills that will allow her to continue with a career later in life.

FIG 11.7.1 Teenage pregnancy is not a good idea

FIG 11.7.2 Babies are expensive

FIG 11.7.3 Being a good mother takes knowledge and patience

Teenage girls do not become pregnant by magic. For every teenage mother, there is a father. Although it is the woman who becomes pregnant and has a baby, the man must also share an equal responsibility.

FIG 11.7.4 Mature mothers can return to well-paid jobs

FIG 11.7.5 Teenage boys need to act responsibly

Many of the reasons why a woman should wait until she is an adult before having a baby apply equally well to a man. A mature man who has completed his education and has a job can provide for his partner and their baby, while an immature father cannot.

Activity 11.7.1

How teenage pregnancy might affect family life

If possible you should work in a mixed group of boys and girls for this activity.

In your mixed group of boys and girls, discuss the effects of a teenage pregnancy on you and your family. Make notes of important points that you can use in a class discussion.

Check your understanding

1. Give as many reasons as you can why it would be far more sensible for a woman to have a baby when she is 25 years old rather than when she is 15 years old.

Review of Embryo development and birth control

- During fertilisation a sperm combines with an egg to form a zygote.

- Fertilisation normally takes place in the oviduct.

- A fertilised egg undergoes rapid cell division.

- The developing baby is initially called an embryo but after 9 weeks it is referred to as a foetus.

- An umbilical cord connects the developing foetus and the placenta on the uterus wall.

- The foetus develops in a sac called the amnion, which is full of watery amniotic fluid.

- The amniotic fluid acts as a shock absorber, preventing the developing foetus from damage.

- Everything that enters and leaves the developing foetus does so through the umbilical cord.

- The mother's blood supply and the foetus's blood supply are not directly joined; substances pass from one to the other by diffusion in the placenta.

- The gestation period is the time taken for a fertilised egg to develop to the point where a baby is born, and in humans this takes 9 months.

- Pre-natal care is care a mother receives whilst she is pregnant, that is before the birth of her baby.

- Expectant mothers attend pre-natal clinics, where their wellbeing is monitored during pregnancy.

- Ultrasound images show an image of the foetus.

- Harmful substances can pass from the mother's blood into the foetus's blood.

- Doctors advise that expectant mothers should not smoke, drink excessive amounts of alcohol or take recreational drugs during pregnancy.

- Viruses can also pass from the mother to the developing foetus.

- In some countries girls are vaccinated against German measles at puberty as contracting this disease during pregnancy can have serious consequences for the developing foetus.

- Contraception is about preventing a woman becoming pregnant as a result of intercourse.

- Natural methods of contraception include coitus interruptus, the rhythm method and abstinence.

- Barrier methods of contraception include the condom, coil/loop and cap/diaphragm.

- Hormonal methods of contraception include the hormone pill.

- Surgical methods of contraception include vasectomy and tubal ligation.

- Teenage women are physically able to have a baby but their bodies are not fully developed and there are many other problems.

- Bringing up a baby is very demanding in terms of time, money and dedication.

- A woman is in a much better position to have a baby when she has finished her education, has worked for several years so she has acquired skills, and is in a stable relationship with a partner.

Review questions on Embryo development and birth control

1. Fig 11.RQ.1 shows a part of the female reproductive system and some stages leading up to pregnancy.

   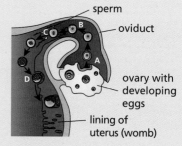
   FIG 11.RQ.1

   a) What process is taking place at:

      i) A?   ii) B?   iii) C?

   b) What happens to the embryo between C and D?

   c) What is happening at D?

2. Fig 11.RQ.2 represents a human foetus in the mother's womb.

   a) Identify parts A–E on the diagram.

   b) Here is a list of things found in blood.

   | glucose   carbon dioxide   oxygen   red blood cells   urea   white blood cells |

   From this list choose two things that pass through part D:

      i) from the mother to the foetus

      ii) from the foetus to the mother.

   c) Describe the function of part A.

   d) What normally happens to part B just before childbirth?

   e) By what process do substances pass between the mother's blood and that of the foetus?

   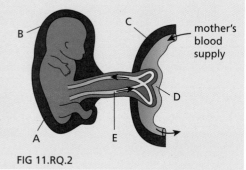
   FIG 11.RQ.2

3. The graph in Fig 11.RQ.3 shows how a typical foetus increases in length as it grows in the mother's uterus.

   a) How are doctors able to measure the size of a growing foetus?

   b) If the foetus is 250 mm long how old is it likely to be?

   c) How big will the baby be at birth if this takes place at 39 weeks?

   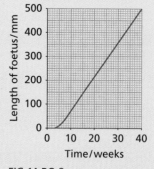
   FIG 11.RQ.3

4. In some countries girls are vaccinated against the rubella virus.

a) What disease is brought about by rubella?

b) Why is it a problem if a pregnant woman contracts this disease?

c) Explain why girls are vaccinated around the time they start to become sexually mature.

d) Explain why boys do not receive this vaccination.

5. State one advantage and one disadvantage of each of the following forms of contraception.

a) Rhythm method

b) Condoms

c) Hormone pill

d) Vasectomy

Improving communication with teenagers

The following is a quote from a website containing information about teenage pregnancy:

> 'Most experts believe that one of the major causes of teenage pregnancy in Jamaica is ignorance regarding proper sexual behaviour and consequences.' (www.pregnancyexposed.com)

FIG 11.SIP.1

Although much information is available for teenagers it seems that many are not getting the important messages.

A new approach has been suggested that involves providing information about teenage pregnancy in the form of a comic strip in school magazines and newsletters. It is thought that some teenagers might respond to this more readily than to formal posters and leaflets.

Here is an example of a comic strip to give you some idea of how one is laid out. Your strip is not meant make people laugh but to deliver an important message. However, if it does contain some element of humour this might give it more credibility with the target audience.

1. You are going to work in groups of 3 or 4 to produce a comic strip suitable for inclusion in your school magazine or newsletter. The tasks are:

   - To review the parts of the unit which are concerned with contraception and teenage pregnancy and make sure you understand all of the issues discussed.

   - To look at examples of cartoon strips in local newspapers and on the Internet to see how they are laid out and put together.

   - To create the first draft of a comic strip that can be reviewed by people from outside your group.

   - To revise your comic strip and produce the final version.

   - To prepare an oral report in which you will explain why you chose a particular focus and discuss the message you believe is delivered by your comic strip. Your report should also describe any particular techniques that you used in its production.

   a) Look back through the unit, and in particular those lessons dealing with different forms of contraception and the problems associated with teenage pregnancy.

   b) Look at examples of comic strips in comics and newspapers. How many separate pictures are in the strip? How has the artist avoided covering the characters with the speech bubbles? Do the images look more effective in colour or in black and white?

c) Before you start thinking about drawing you need a story line. What message is your comic going to deliver? A simple message well delivered is likely to have a much greater impact than a complicated message that few readers will understand.

d) Once you have decided on your story line you need to assess how many pictures you will need to deliver it. You might be able to deliver it in one picture or you might need a sequence of pictures that follow on. You would be wise to limit this to three or if really necessary, to four. You are seeking to deliver a message, not create a soap opera.

FIG 11.SIP.2 A picture drawn on a computer

e) How are you going to draw your pictures? Cartoonists often draw pictures in a larger scale than the final version. This allows them to add detail more easily. When a picture is finished it can be reduced by a photocopier, or scanned into a computer, reduced and reprinted.

Does your school have any computer-aided drawing packages or other software that would allow you to create your images using a computer? Can you make use of any clip art in your designs? You might be able to drop images of people into a computer-drawn picture.

If you decide to draw freehand don't worry about adding colour at this stage. The outlines and text will provide you with a version that can be reviewed by people outside your group.

f) Who are you going to ask to review your work? You might ask your science teacher about the content and your art teacher about the design, but at the end of the day the message you are planning to deliver is directed towards students around your own age. You need to make sure that what you have done makes sense to them so be sure to include students in your team of reviewers.

g) Once your first version has been reviewed it is time to make whatever changes are necessary. Once you are satisfied with the content carry on and produce the final version of your comic strip.

h) Your audience will need to have access to your comic strip when you come to give your oral presentation. This might be by printing copies for circulation in advance or it might be by projecting an image onto a screen.

If your comic strip is successful in delivering its message you should not need to explain what this is to your audience. You should explain why you believe this message is important and how you think it will help to reduce teenage pregnancy.

Your audience might also be interested in how you went about creating your comic strip, so you should be prepared to describe and discuss the different stages. Do not be reluctant to describe things that went wrong as learning through mistakes is an important aspect of scientific development.

Index

Note: Page numbers followed by *f* or *t* represent figures or tables respectively.

A

acceleration, 13

accuracy, measurement, 18

acid–alkali indicators, 39, 184–185, 192, 193

acid–alkali reactions, 186–187

acid–carbonate reactions, 195, 195*f*

acids, 174–175, 174*f*, 175*f*, 187

    defined, 177

    in fruits, 177

    identification of, 176–177

    neutralisation of, 187, 188–189

    reactions of, 194–195

    strength of, 178–179, 179*f*

ADH. *see* antidiuretic hormone (ADH)

adrenal glands, 164*f*, 165

adrenalin, 166, 166*f*, 167

agar gel, 38, 38*f*, 39

alkalis, 174–175, 174*f*, 187

    defined, 181

    identification of, 180–181

    neutralisation of, 187, 188–189

    strength of, 182–183, 183*f*

alloys, 109, 109*f*

ammeter, 88, 88*f*, 89

ammonia, 192–193

ammonia solution, 181, 181*f*, 192, 192*f*

ammonium hydroxide, 181

amnion, 206, 207

amniotic fluid, 206, 207

ampere (amp, symbol A), 88, 89

analogue ammeter, 88, 88*f*

anions, 134, 135, 152

antacids, 190, 191

antidiuretic hormone (ADH), 166, 167

antistatic, 72

ants, 183

arteries, 40*f*, 41, 44, 44*f*, 45, 48

arterioles, 45

artificial heart, 52–53, 52*f*, 53*f*

atoms, 68, 69, 70, 134

    carbon, model of, 69*f*

    representation of, 130–131, 130*f*

    sodium, structure of, 134*f*

    structure of, 68*f*

auditory nerve, 156, 157

auricles (atria), 42, 42*f*, 43

B

baking powder, 190, 190*f*, 191

bar magnets, 110*f*

battery, 84, 85

    car, 87

    and circuit symbols, 91, 91*f*

    fruit, 85, 85*f*

beam balance, 10

begonias (plant), 55

binary compound, 132, 133

birth control, 210, 211

blinking, 160

blood

    components of, 46–47

    concentration of glucose in, 167, 167*f*

    deoxygenated, 41, 41*f*, 42, 43

    flow through heart, 42

    oxygenated, 41, 41*f*, 42, 43

blood vessels, 168, 168*f*

bonding, ionic compounds, 152–153

brain, 158–159, 160

    cerebellum, 158, 158*f*

    cerebrum, 158, 158*f*

    medulla oblongata, 158*f*, 159

brain stem, 159

C

calcium hydroxide, 175

calcium sulfate, 139, 175, 175*f*

capillaries, 45, 45*f*

capillarity, 60, 61

car battery, 87

carbon, 68*f*

    atom, model of, 69*f*

carbonate, 194, 195

carbon dioxide, 190, 191, 194

carnivorous plants, 63, 63f

cations, 134, 135, 152

cell division, 206, 206f, 207

cells (electrical), 84, 85
    connection in parallel, 97, 97f
    connection in series, 95, 95f
    and lamps, 86–87, 86f, 87f

cells, transport across, 26–32
    diffusion process, 26–27, 27f
    osmosis. see osmosis

Celsius (Centigrade) (°C), 9

central nervous system (CNS), 154, 155, 155f

ceramic magnets, 109, 109f

cerebellum, 158, 158f

cerebrum, 158, 158f

charcoal, 16, 16f

chemical equations, 136, 137, 144

chemical formula
    of compound, 130
    of elements, 130

chemical hand warmers, 149

chemical messengers. see hormones

chlorophyll, 55

circuit breaker, 121, 121f

circuit diagram, 90, 90f, 91
    constructing circuits from, 92–93, 92f

circuits
    complete, 84–85, 84f
    construction from circuit diagrams, 92–93, 92f
    defined, 84, 85
    parallel, connecting components in, 96–97, 96f, 97f
    series, connecting components in, 94–95, 94f, 95f
    symbols, 90–91, 90f, 91f
    voltage in, 98

circulatory system
    arteries, 40f, 41, 44, 44f, 45
    capillaries, 45, 45f
    double, 40–41, 40f
    heart, 40, 41, 42–43, 42f
    structure of, 40–41, 40f
    veins, 40f, 41, 44, 44f, 45

citric acid, in dyeing, 200–201

CNS. see central nervous system (CNS)

coil/loop, 211t

combining power, 132–133, 132t, 133f

combustion, 148, 148f, 149

combustion reaction, 137

communication, with teenagers, 220–221

compass needle, 111, 111f

complete circuit, 84–85, 84f

components
    in electrical circuit diagram, 90, 90f, 91
    in parallel, 96–97, 96f, 97f
    in series, 94–95, 94f, 95f

compounds
    ammonium, 192–193
    binary, 132, 133
    chemical, solubility of, 146, 146t
    chemical formula of, 130
    defined, 130, 131
    ionic, 138–139, 152

concentration, 27

concentration gradient, 27

condom, 211, 211f

conductors, 70, 71, 80, 81

contraception, 210–213
    barrier methods, 210, 211t
    hormonal methods, 212, 212t, 213
    natural methods, 210, 210t
    surgical methods, 212, 213t

control experiment, 14

control variables, 14

conventional current flow, 84, 84f, 85

credit cards, 119, 119f

cup/diaphragm, 211, 211t

current
    conventional flow, 84, 84f, 85
    induced, 122–123, 122f
    magnetic effect of, 114–115, 114f, 115f
    measurement, 88–89
    and voltage, 98

D

data, 10

debit cards, 119, 119f

density/dense, 192, 193

deoxygenated blood, 41, 41f, 42, 43

Index

dependent variables, 14
diabetes, 167
diastole, 42–43
differentially permeable membrane, 28, 28f, 29
diffusion, 26–27, 27f, 204, 205
   osmosis *vs.*, 32–33, 33t
   and surface area, 38–39
digital ammeter, 88, 88f
displacement reaction, 147
distance-time graph, 12, 12f, 13
documents, legibility of, 172–173
double circulatory system, 40–41, 40f
double insulated, 83
duplet, 140, 141
dyeing, citric acid in, 200–201
dynamic equilibrium, 32, 33

E

ears, 156, 156f
earth wire, 83, 83f
effectors, 154, 155
egg cell, 202, 202f
electrical appliances, 83, 83f, 89
electrical conductors. *see* conductors
electrical insulators. *see* insulators
electric bell, 120, 120f
electric current. *see* current
electricity, 80
   and safety, 82–83
   static. *see* static electricity
electromagnetism, 117
electromagnets, 116–117, 117f
   strength of, 118–119
   uses of, 120–121
electromotive force, 122, 123
electron flow, 84, 84f, 85
electronic balance, 8
electronic configuration
   defined, 140, 141
   of elements, 140, 140t
   of ions, 141, 141f
electrons, 68, 68t, 69, 81, 134, 140
   mobile, 138
electroscope, building, 78–79, 78f, 79f
electrostatic, 72

electrostatic charging
   photocopiers, 72–73, 72f
   spray painting, 72, 72f
elements, 68, 68f, 69
   chemical formula of, 130
   combining power, 132–133, 132t, 133f
   electronic configuration of, 140, 140t
   metallic, 130
   molecule of, 130
   non-metallic, 130
embryo, 202, 203
encoding machine, building, 106–107, 106f, 107f
endocrine system, 164–165, 164f
   nervous system *vs.*, 168–169
endothermic reactions, 148, 148f–149f, 149
equations
   balancing, 144–145
   chemical, 136, 137
   ionic, precipitation reactions and, 146–147
   symbol, 136, 144–145
   synthesis reaction, 136, 137
   word, 136–137
equilibrium, 28, 29
evaporation, 63
executive toy, creation using magnets, 128–129, 128f, 129f
exercise, and pulse rate, 48–49
exothermic reactions, 148, 148f, 149
experiment
   control, 14
   designing, 16–17
   on osmosis, 30–31
   structuring, 14–15
eyes, 156, 156f

F

fair test, 14, 30, 31
fertilisation, 202–203, 202f
fibrous root, 56, 56f, 57
filtrate, 146, 147
flick book, 25, 25f
flowers, 54f, 55
foetus, 203
   development of, 206–207, 206f, 207f
friction, 70, 71, 71t, 148, 148f, 149
fruit battery, 85, 85f

G

galvanometer, 122, 122*f*

responses on, 122*t*

gestation period, 206, 207

giraffe, 48

glands

adrenal, 164*f*, 165

pituitary, 164, 164*f*, 165

thyroid, 164, 164*f*, 165

glucagon, 166, 167

gold-leaf electroscope, 78, 78*f*

gradient, 12, 13

concentration, 27

graph, 12

distance-time, 12, 12*f*, 13

velocity-time, 12–13, 12*f*

H

heart, 40, 41, 42–43, 42*f*

artificial, making, 52–53, 52*f*, 53*f*

chambers of, 42, 42*f*

pumps, 169

heat energy, 16, 16*f*

HIV, 209

hormonal methods, 212, 212*t*, 213

hormone pill, 212*f*, 212*t*

hormones, 164, 164*f*, 165,
166–167, 169

produced by pituitary gland, 166

horseshoe magnet, 110*f*

hydrogen, 195

hydroxide, 181

hypothesis, 14, 30, 31

I

independent variables, 14

indicators, 177

acid-alkali, 184–185

universal, 179, 179*f*, 183, 183*f*

indigestion, 187

indirect method, temperature measurement,
16, 17

induced current, 122–123, 122*f*

insulators, 70, 71, 80, 81. *see* electrical
insulators

insulin, 166, 167

involuntary action, 160, 161

ionic compounds, 152

bonding, 152–153

defined, 138, 139

melting point, 138, 138*t*

properties of, 138–139

solubility, 138

'sparingly soluble,' 139

ionic equations, precipitation reactions and,
146–147

ions, 134–135, 134*t*, 138, 138*f*

electronic configuration of, 141, 141*f*

sodium, 152

isoelectronic, 141

K

Kelvin (K), 9

knee-jerk reflex, 160–161

L

lamps

cells and, 86–87, 86*f*, 87*f*

in parallel, 96–97, 96*f*, 97*f*

in series, 95, 95*f*

symbols for, 91, 91*f*

law of conservation of mass, 142–143

leaves, 54*f*, 55

lightning, 70, 70*f*, 71, 74

lightning conductors, 74, 74*f*, 75

like poles, 111

liquid-in-glass thermometer, 9, 9*f*

litmus, 184, 184*f*, 185

lodestone, 108, 108*f*

lumen, 44, 45

lungs, 38, 38*f*, 39

lymphocytes, 47

M

magnesium, burning in air, 144, 144*f*

magnetic effect of current, 114–115, 114*f*,
115*f*

magnetic field lines, 112, 113*f*

for unlike poles, 113, 113*f*

magnetic fields, 112–113, 113*f*

magnetic field strength, 113

magnetic football, 128, 128*f*

magnetic materials, 109

magnetite, 108

Index

magnets
    bar, 110*f*
    ceramic, 109, 109*f*
    executive toy creation using, 128–129, 128*f*, 129*f*
    game creation using, 128–129, 128*f*, 129*f*
    horseshoe, 110*f*
    magnetic field around, 112–113, 113*f*
    permanent, 109
    poles, 110–111, 110*f*
mass, 8, 143
    law of conservation, 142–143
matrix, 138, 139
medulla oblongata, 158*f*, 159
melting point, ionic compounds, 138, 138*t*, 139
menstrual cycle, 203
metallic elements, 130
metallic structure, 81, 81*f*
metalloids, 68, 68*f*
metals, 68, 68*f*
    structure of, 81, 81*f*
mobile electrons, 138, 139
molecules
    defined, 130, 131
    representation of, 130–131
multimeter, 88, 89

N
nerve cells, 168, 168*f*
nervous system, 154, 154*f*
    central, 154, 155, 155*f*
    endocrine system *vs.*, 168–169
    peripheral, 154, 155, 155*f*
    and reaction time, 162, 162*f*
neurons, 168, 168*f*
neutralisation, 187, 188–189, 188*f*
    product of, 189
    reactions in everyday life, 190–191
neutralisation reaction, 145
neutrons, 68, 68*t*, 69
nichrome wire, 89
non-magnetic materials, 109
non-metallic elements, 130
non-metals, 68, 68*f*
nose, 156, 156*f*

O
observations
    qualitative, 10
    quantitative, 10–11, 10*f*
octet, 140, 141
oestrogen, 166, 167
olfactory nerve, 156, 157
optic nerve, 156, 157
organs
    sense, 156–157, 156*f*
    target, 164, 165
osmosis, 28–29
    diffusion *vs.*, 32–33, 33*t*
    planning experiment on, 30–31
    reverse, 36–37, 36*f*, 37*f*
ovaries, 164*f*, 165
oviducts, 202, 202*f*
oxidation reaction, 137
oxygenated blood, 40, 40*f*, 42, 43

P
pancreas, 164, 164*f*, 165, 166
parallel, connecting components in, 96–97, 96*f*, 97*f*
peripheral nervous system (PNS), 154, 155, 155*f*
permanent magnets, 109
    uses of, 120–121
pH, 179
phagocytes, 47
phenolphthalein, 39, 39*f*
phloem, 58–59, 58*f*
    structure of, 60, 60*f*
photocopiers, 72–73, 72*f*
photosynthesis, 148, 148*f*, 149
physical quantity, 8
pituitary gland, 164, 164*f*, 165
    hormones produced by, 166
placenta, 204–205, 204*f*
placentophagy, 205
plants
    carnivorous, 62, 62*f*
    flowers, 54*f*, 55
    leaves, 54*f*, 55
    movement of substances, 60
    parts of, 54, 54*f*
    petals, 184, 184*f*

phloem, 58–59, 58*f*, 60, 60*f*

  pot, 66–67, 66*f*, 67*f*

  roots. *see* roots

  stem, 54, 54*f*, 55

  transpiration through, 62–63, 62*f*

  xylem, 58–59, 58*f*, 60, 60*f*

plasma, 46

platelets, 46, 47

PNS. *see* peripheral nervous system (PNS)

poles, magnets, 110–111, 110*f*

polyatomic ions, 135

polythene manufacture, 74–75, 74*f*

potential difference, 87, 98, 99

pot plants, 66–67, 66*f*, 67*f*

powders, in pipes, 74

power, combining, 132–133, 132*t*, 133*f*

precipitate, defined, 146, 147

precipitation reactions, 142, 143

  and ionic equations, 146–147

prefixes, 7, 7*t*

pregnancy

  foetus, development of, 206–207, 206*f*, 207*f*

  pre-natal care, 208–209

  smoking during, 208, 208*f*

  teenage, 214–215, 214*f*

pre-natal care, 208–209

products, 136, 137

progesterone, 166

protons, 68, 68*t*, 69

pulse rate, 48, 49

  exercise and, 48–49

pulses, 48, 48*f*

pungent, 192, 193

Q

qualitative observations, 10

quantitative observations, 10–11, 10*f*

quantity

  measurement. *see* quantity measurement

  physical, 8

quantity measurement, 8–9

  mass, 8

  temperature, 9, 9*f*

  time, 8

  volume, 8

R

radial pulse, 48, 49

radioactive isotope, 75

reactants, 136, 137

reaction time, 162, 162*f*

receptors, 154, 155

red blood cells, 46, 46*f*, 47

reflex action, 160–161

reflex arc, 160, 161, 161*f*

relay, 121, 121*f*

residue, 146, 147, 189

reverse osmosis, 36–37, 36*f*, 37*f*

root hair, 56*f*, 57

roots, 54, 54*f*, 55

  cells, 56*f*, 57

  fibrous, 56, 56*f*, 57

  tap, 56, 56*f*, 57

rubella, 208, 209

S

safety, 196–197

  electricity and, 82–83

salt, 152, 152*f*, 189

scientific notation, 21

sense organs, 156–157, 156*f*

septum, 42, 43

series, connecting components in, 94–95, 94*f*, 95*f*

sherbet fizz, 191, 191*f*

sieve elements, 60, 60*f*

sieve tube, 60, 61

significant figures, 20, 20*f*

silver, 81

silver iodide, 146, 146*f*

SI units, 6, 6*t*, 7

skin, 156–157, 156*f*

skull, 155

small intestine, 38, 39

smoking, during pregnancy, 208, 208*f*

sockets, 82, 82*f*

sodium ions, 152

solenoid, 115, 115*f*, 121

solubility

  chemical compound, 146, 146*t*, 147

  ionic compounds, 138

'sparingly soluble' ionic compounds, 139

sperm, 202, 202f

spermicidal creams, 212t

spray painting, 72, 72f

standard form, 21

static, meaning of, 70, 71

static electricity, 70, 70f

    charging with, 70, 71

    friction and, 70, 71, 71t

    problems caused by, 74–75

    uses of, 72–73

STEAM (Science, Technology, Engineering, Art and Mathematics) activities, 24–25

stem, plants, 54, 54f, 55

    vascular bundles, 58, 58f, 59

stimulus, 154, 155

Sturgeon, William, 117

sub-atomic particles, 68, 68t, 69

sulfuric acid, 175, 175f

surface area, diffusion and, 38–39

symbol equations, 136

    balancing, 144–145

symbols, circuits, 90–91, 90f, 91f

systole, 43

T

tap root, 56, 56f, 57

target organ, 164, 165

tea, 33

teenage pregnancy, 214–215, 214f

teenagers, communication with, 220–221

temperature, 9, 9f

template, 16, 17

testes, 164f, 165

testosterone, 166, 167

thermometer, liquid-in-glass, 9, 9f

thyroid gland, 164, 164f, 165

thyroxine, 166, 167

time measurement, 8

titration, 188f, 189

tongue, 156, 156f

transpiration, 62–63, 62f

tubal ligation, 213, 213f, 213t

U

ultrasound, 208, 208f, 209

umbilical cord, 204, 204f, 205

universal indicator, 179, 179f, 183, 183f

unlike poles, 111

    magnetic fields for, 113, 113f

V

variables

    control, 14

    dependent, 14

    independent, 14

vascular bundles, 58, 58f, 59

vascular cambium, 58, 59

vasectomy, 213, 213f, 213t

veins, 40f, 41, 44, 44f, 45

velocity-time graph, 12–13, 12f

ventricles, 42, 42f, 43

venules, 45

Venus flytrap, 63, 63f

voltage

    current and, 98

    defined, 98, 99

    measurement, 98

voltmeter, 98, 98f, 99

volts (V), 87

volume, 8

voluntary action, 160, 161

W

white blood cells, 46, 46f, 47

word equations, 136–137

X

xylem, 58–59, 58f

    structure of, 60, 60f

Z

zygote, 202, 203, 206

Acknowledgements

The publishers wish to thank the following for permission to reproduce photographs. Every effort has been made to trace copyright holders and to obtain their permission for the use of copyright materials. The publishers will gladly receive any information enabling them to rectify any error or omission at the first opportunity.

p6–7: algabafoto/Shutterstock, p6: Foodfolio/Alamy Stock Photo, p6: Carolyn Jenkins/ Alamy Stock Photo, p6: Joey Chung/Shutterstock, p7: AlenKadr/Shutterstock, p8: Science Photo Library/Getty Images, p8: Llitchima/Shutterstock, p8: Ziviani/Shutterstock, p9: ID1974/ Shutterstock, p9: Lipskiy/Shutterstock, p10: ANDREW LAMBERT PHOTOGRAPHY/SCIENCE PHOTO LIBRARY, p10: PRILL/Shutterstock, p10: CHARLES D. WINTERS/SCIENCE PHOTO LIBRARY, p15: Al Freni/Getty Images, p15: Anucha Cheechang/Shutterstock, p16: Jag_cz/ Shutterstock, p18: smikeymikey1/Shutterstock, p18: pzAxe/Shutterstock, p20: windu/ Shutterstock, p20: moprea/Shutterstock, p24: MyLoupe/Getty Images, p25: Iosw/Shutterstock, p25: ANDRE DURAND/Getty Images, p26–27: Kateryna Kon/Shutterstock, p26: Testing/ Shutterstock, p27: SCIENCE PHOTO LIBRARY, p30: Alexandre Dotta/SCIENCE PHOTO LIBRARY, p32: gresei/Shutterstock, p36: David Litman/Shutterstock, p38–39: Kateryna Kon/ Shutterstock, p38: sruilk/Shutterstock, p39: Alexeysun/Shutterstock, p47: Steve Debenport/ Getty Images, p48: Joel Shawn/Shutterstock, p49: John Cole/Science Photo Library, p54–55: Barbol/Shutterstock, p56: Rattiya Thongdumhyu/Shutterstock, p56: Richard Griffin/Shutterstock, p61: Veniamin Kraskov/Shutterstock, p62: CORDELIA MOLLOY/SCIENCE PHOTO LIBRARY, p63: Studio Barcelona/Shutterstock, p66: ROBYN BECK/AFP/Getty Images, p67: Rigamondis/ Shutterstock, p68–69: Bildagentur Zoonar GmbH/Shutterstock, p68: Ratchat/Shutterstock, p68: Hywit Dimyadi/Shutterstock, p68: Maurice Savage/Alamy Stock Photo, p70: Borges Samuel/ Alamy Stock Photo, p70: Maryna Patzen/Shutterstock, p70: Dorling Kindersley/UIG/SCIENCE PHOTO LIBRARY, p71: MARTYN F. CHILLMAID/SCIENCE PHOTO LIBRARY, p74: Alamy stock Photo, p74: Wesley/Keystone/Getty Images, p75: Slavko Sereda/Shutterstock, p78: SCIENCE PHOTO LIBRARY, p80–81: Wang An Qi/Shutterstock, p82: PhotoSerg/Shutterstock, p82: PhotoSerg/Shutterstock, p83: John Kasawa/Shutterstock, p83: Krivoshein Igor Alexandrovich/ Shutterstock, p83: sue yassin/Shutterstock, p84: Paddington/Shutterstock, p88: sciencephotos/ Alamy Stock Photo, p88: Prasolov Alexei/Shutterstock, p98: Tanasan Sungkaew/Shutterstock, p98: Dmitrii Kazitsyn/Shutterstock, p106: Lenscap Photography/Shutterstock, p108–109: piick/ Shutterstock, p108: Phil Degginger/Alamy Stock Photo, p109: Fedorov Oleksiy/Shutterstock, p109: Peter Sobolev/Shutterstock, p111: terekhov igor/Shutterstock, p113: CORDELIA MOLLOY/ SCIENCE PHOTO LIBRARY, p117: Anton Kozlovsky/Shutterstock, p117: Science & Society Picture Library/Getty Images, p117: Petar An/Shutterstock, p119: Gomolach/iStockphoto, p121: RomanStrela/Shutterstock, p122: charistoone-stock/Alamy Stock Photo, p128: 1ABilder/ Alamy Stock Photo, p128: iMoved Studio/Shutterstock, p128: Iain McGillivray/Shutterstock, p128: Elena Schweitzer/Shutterstock, p129: NikolayN/Shutterstock, p130–131: Elaine Barker/ Shutterstock, p136: g_tech/Shutterstock, p138: GIPhotoStock/SCIENCE PHOTO LIBRARY, p144: sciencephotos/Alamy Stock Photo, p144: CHARLES D. WINTERS/SCIENCE PHOTO LIBRARY, p146: MARTYN F. CHILLMAID/SCIENCE PHOTO LIBRARY, p148: Vitalina Rybakova/ Shutterstock, p148: Kunertus/Shutterstock, p148: Elenamiv/Shutterstock, p149: Carolyn Jenkins/ Alamy Stock Photo, p152: Evlakhov Valeriy/Shutterstock, p154–155: adike/Shutterstock, p154: wavebreakmedia/Shutterstock, p154: Peter Titmuss/Alamy Stock Photo, p156: Dan Kosmayer/Shutterstock, p156: PhotoMediaGroup/Shutterstock, p156: schankz/Shutterstock, p156: Baerbel Schmidt/Getty Images, p156: Matthias G. Ziegler/Shutterstock, p158: JGI/Tom Grill/Blend Images/Alamy Stock Photo, p158: CREATISTA/Shutterstock, p159: Wendy Hope/ Getty Images, p160: Anthony Asael/Art in All of Us/Contributor/Getty Images, p160: Sean De Burca/Getty Images, p160: AdamEdwards/Shutterstock, p162: Hogan Imaging/Shutterstock,

Acknowledgements

p162: michaeljung/Shutterstock, p162: DIBYANGSHU SARKAR/AFP/Getty Images, p162: mHanafiAhmad/Shutterstock, p166: Tino Soriano/Getty Images, p166: Douglas Pearson/Getty Images, p167: Science Photo Library/Alamy Stock Photo, p168: Romanova Natali/Shutterstock, p168: Inok/Getty Images, p172: Blend Images/Shutterstock, p174–175: PowerUp/Shutterstock, p174: Svitlana-ua/Shutterstock, p174: Gts/Shutterstock, p175: Dr P. Marazzi/Science Photo Library, p175: stockphoto mania/Shutterstock, p178: Coprid/Shutterstock, p178: MARTYN F. CHILLMAID/SCIENCE PHOTO LIBRARY, p178: Pulse/Fuse/Getty Images, p179: Charles D. Winters/Science Photo Library, p181: Giphotostock/Science Photo Library, p183: ANDREW LAMBERT PHOTOGRAPHY/SCIENCE PHOTO LIBRARY, p184: Henrik Larsson/Shutterstock, p184: Subbotina Anna/Shutterstock, p184: artphotoclub/Shutterstock, p184: bergamont/Shutterstock, p189: Andrew Lambert Photography/Science Photo Library, p190: Whitebox Media/Alamy Stock Photo, p190: hlphoto/Shutterstock, p190: absolutimages/Shutterstock, p190: Shahril KHMD/Shutterstock, p191: Milleflore Images/Shutterstock, p192: ANDREW LAMBERT PHOTOGRAPHY/SCIENCE PHOTO LIBRARY, p193: geogphotos/Alamy Stock Photo, p194: Andrew Lambert Photography/Science Photo Library, p194: Martyn F. Chillmaid/Science Photo Library, p195: Charles D. Winters/Science Photo Library, p196: Andreas von Einsiedel/Alamy Stock Photo, p197: Klaus Vedfelt/DigitalVision/Getty Images, p197: Huntstock, Inc/Alamy Stock Photo, p197: Olga Kovalenko/Shutterstock, p200: bubutu/Shutterstock, p201: Jiri Hera/Shutterstock, p201: fototrips/Shutterstock, p201: AnnapolisStudios/Shutterstock, p202–203: zffoto/Shutterstock, p204: crabgarden/Shutterstock, p204: Casa nayafana/Shutterstock, p208: Monkey Business Images/Getty Images, p208: Chad Ehlers/Stock Connection/Getty Images, p208: photorevolution/Getty Images, p209: AUBERT/BSIP/Alamy Stock Photo, p209: Tatiana Chekryzhova/Shutterstock, p212: toons17/Shutterstock, p212: SSPL/Getty Images, p214: Samuel Borges Photography/Getty Images, p214: adventtr/Getty Images, p214: michel74100/Getty Images, p215: Ivonne Wierink/Shutterstock, p215: iofoto/Shutterstock, p218: Denis Cristo/Shutterstock.